Issues in healthcare risk management

Edited by Stuart Emslie and Charles P Hancock

Copyright © 2008 Editors and contributors

Published by:
Healthcare Governance Limited,
32 Harpes Road,
Oxford,
OX2 7QL
United Kingdom
www.healthcaregovernance.com

ISBN 978-0-9558526-0-2

Contents

Foreword

Acknowledgments

Editors and contributors

Preface

1. Consideration of the essential interacting elements of a risk management system
   JANE RIPPON

2. A review of the extent to which the Trust Risk Management Strategy is functioning within the Emergency Services Division of an NHS Trust
   CAROLE MODERATE

3. The application of a risk management approach in the planning, development and commissioning of an NHS Walk-in Centre
   LINDA CAMP

4. Implementation of a risk management strategy within the Chronic Disease Management Team of a Primary Care Trust
   EMILY HACKETT

5. Reducing claims against the NHS through the rapid and sensitive handling of complaints - I
   ROBERT CALDEIRA

6. Reducing claims against the NHS through the rapid and sensitive handling of complaints - II
   JAYNE HARTLEY

7. The relationship between hospital acquired infection rates and the contracting out of cleaning services in the NHS in England - I
   KIM HUDSON

8. The relationship between hospital acquired infection rates and the contracting out of cleaning services in the NHS in England - II
JAYNE HARTLEY

9. Reducing the frequency and impact of needlestick injuries involving healthcare staff
JAYNE HARTLEY

10. The ageing National Health Service workforce: A significant risk to the NHS and to the nation?
CAROLE MODERATE

11. Violence and aggression towards health care staff - I
CAROLE MODERATE

12. Violence and aggression towards health care staff - II
BECKY MONAGHAN

13. Fire safety and the training of staff in fire prevention and management in healthcare premises
JAYNE HARTLEY

14. The risks and opportunities presented to the NHS by the disposal of surplus buildings
SHIRLEY MUNDAY

Foreword

The delivery of effective healthcare involves the management of risk at every level of analysis. There are individual, organisational, and population level risks. There are risks in the delivery and operation of providers of healthcare as well as strategic planning and political risks. These risks will never be eliminated but by recognition of the range of issues that affect the quality of care that any one individual may receive, we can optimise the outcomes. The aim is therefore to arrive at a careful analysis of the issues and the factors that affect clinical outcomes so that we can move to a position where the full range of risks are mitigated or 'managed'.

The publication in UK of *An Organisation with a Memory* by the Chief Medical officer, Professor Sir Liam Donaldson, provoked a real interest in the problems of patient safety. It exposed the lack of knowledge in the NHS of the causes and sources of risk in many areas of patient management. The identification and management of risk became an area of growing research and clinical debate. The National Patient Safety Agency was established and both patients and staff alike became more aware of the costs as well as the benefits of our healthcare system.

But to the concern of all, the incidence of adverse events still remains high. This book takes a novel approach to the problem and reports a range of investigations into healthcare risk that are conducted by senior and committed members of NHS staff who also have an academic training. This combination of 'front line' experience and academic investigative rigour leads to an interesting analysis of a range of different aspects of risk in healthcare delivery.

Loughborough University has a strong tradition of applied research and employer oriented teaching and research programmes and this book is firmly in that tradition. It epitomises the benefits of integrating the academic analysis with the practitioner knowledge and skills. In doing so it provides us with an analysis of a very wide range of factors that impact on quality of provision of healthcare. It is to be hoped that it will be widely read across the healthcare world and that other areas of risk will also be subjected to this well informed analysis.

Professor Shirley Pearce CBE
Vice Chancellor
Loughborough University

Acknowledgments

The Editors wish to thank the individual contributors and, where appropriate, their employing organisations for agreeing that their material could be included in this publication.

Thanks are also due to Dr Stuart Evans and John Step who each reviewed various aspects of the draft manuscript.

Editors and contributors

EDITORS

Stuart Emslie

Stuart is an independent governance and risk specialist and Visiting Fellow in governance and risk at Loughborough University Business School, where he leads a postgraduate programme on healthcare governance. He lectures and consults internationally on governance and risk management in healthcare and is formerly Head of Controls Assurance for the NHS in England at the Department of Health.

Charles Hancock

Charles is Director of Healthcare Programmes at Loughborough University Business School, and is responsible for all programmes including postgraduate programmes in healthcare risk management, healthcare governance, backcare management and fire safety. Charles has both a clinical and educational background.

CONTRIBUTORS

The contributors are all senior managers and professionals in the National Health Service who have undertaken the postgraduate programme in healthcare risk management at Loughborough University. This book is an edited collection of some of their assignments. The contributors are: **Robert Caldeira, Linda Camp, Emily Hackett, Jayne Hartley, Kim Hudson, Carole Moderate, Becky Monaghan, Shirley Munday and Jane Rippon.**

Preface

Introduction

Healthcare is an increasingly complex and cost-constrained undertaking, fraught with risk. Risks to patients. Risks to staff. Risks to the public. And risks to the corporate healthcare organisation established as the infrastructure within which modern care is provided. On this basis, contemporary healthcare risk management is not about 'clinical' vs. 'non-clinical' risk. It is about taking a holistic, 'enterprise-wide' approach to risk identification and management. It is about engaging everyone in the process, from front-line staff up to the board. Successfully managing risk is, therefore, a key imperative for the modern healthcare professional, manager and board member.

This is a book with a difference. Most books of this nature would be an edited collection of the great and the good in academia, in high level policy making, and in the NHS and wider healthcare. This book, however, is an edited collection of the works of senior NHS professionals who, in 'later life', decided to embark on a part-time postgraduate course in healthcare risk management at Loughborough University. Uniquely, this book provides insights and perspectives that could only come from a combination of years of practical experience coupled with a fresh enthusiasm for academic pursuits. As a matter of principle, therefore, the Editors have sought to minimise editing in an attempt to preserve the originality of the student work.

Chapter outlines

Chapters 1-4 are concerned with matters of risk management strategy. The remaining chapters, 5-14, deal with more operational aspects of managing risk: claims reduction through better complaints handling; hospital acquired infection and cleaning services; needlestick injuries to staff; the ageing NHS workforce; violence and aggression towards staff; fire safety and fire training for staff; and disposal of surplus NHS buildings.

In chapter 1, Jane Rippon seeks to identify the elements of a risk management system and discuss the impact that the interaction between these elements has on the risk management system within a 650 bed NHS hospital trust. The elements of a risk management system explored in the chapter are those identified by the Australian/New Zealand risk management standard 4360, which has been adopted by all UK health departments (although with the demise of the NHS controls assurance project, Department of Health sponsorship of the Standard appears to have lapsed in England) and is widely used by NHS organisations across the UK.

In chapter 2, Carole Moderate reports on a snapshot review of the functioning of a Trust risk management strategy in the emergency services division of a 400-bed district general hospital in England. She finds that the effectiveness of the strategy can be established through assessing staff perception of risk and risk management within a local clinical area. Lack of effectiveness of the risk management strategy is attributed to ineffective communication and lack of ownership. However, the process of review itself was found to have the benefit of improving awareness amongst staff of the trust risk management strategy.

Linda Camp looks, in chapter 3, at the application of a risk management approach in the planning, development and commissioning of an NHS Walk-in Centre. Walk-in Centres are a relatively recent phenomenon in the history of the development of the NHS, offering nurse-led, no appointment necessary, fast and convenient access to a range of NHS services from early morning to late evening, seven days a week. Linda finds the application of the local Primary Care Trust (PCT) risk management strategy in relation to meeting the objectives of the Walk-in Centre project helped ensure the success of the initiative and, in the process, created a risk aware culture in the PCT.

In the final chapter covering matters of strategy, Emily Hackett looks, in chapter 4, at the effectiveness of the risk management strategy within a service delivery area – the Chronic Disease Management Team - of an NHS Primary Care Trust (PCT). She examines how the strategy is communicated and applied, and identifies weaknesses in the systems described within the strategy that can be improved to strengthen the risk management process as a whole.

Chapters 5 and 6 explore the issue of reducing claims against the NHS through the rapid and sensitive handling of complaints. Robert Caldeira and Jayne Hartley take different perspectives. Caldeira concludes that in looking at the evidence, communication plays an important role in stopping complaints and claims, and in stopping complaints progressing to claims. He believes that the statistical evidence appears to support the hypothesis that rapid and sensitive handling of complaints results in a reduction in claims. Hartley, on the other hand, concludes from her analysis that there appears to be very little published evidence that complaints management has a significant impact on litigation. She does, however, concede that the development of risk management systems which address complaints management, incident reporting, education and training of staff, communication and information provision in conjunction with the civil justice reforms may all have an impact on the incidence of claims.

Taken together, the two chapters provide some thought provoking insights into complaints and claims in the NHS.

The contentious issue of the relationship between hospital acquired infection rates and the contracting out of cleaning services in the NHS in England is the principal subject of chapters 7 and 8. Both authors provide perspectives on hospital acquired infection and its relationship to contract cleaning and other factors. In chapter 7, Kim Hudson finds a tangled web of ambiguous reports and other information relating to the subject and supports the view that it is more than simple cleanliness that impacts on infection prevention. She concludes that actions rather than words are needed to bring about improvements in infection control. Jayne Hartley, in chapter 8, finds that she is unable to show that increased rates of hospital acquired infections have been as a result of the introduction of contracting out of cleaning services in the NHS since 1983. She believes that there are many factors affecting the incidence of hospital acquired infection and there needs to be a shared and collective responsibility to addressing the problem. She concludes by quoting the Department of Heath in saying that "keeping the NHS clean is everybody's responsibility."

Needlestick injury to staff is the topic of chapter 9. In this chapter, Jayne Hartley looks at reducing the frequency and impact of needlestick injuries to healthcare staff through implementing appropriate risk management strategies that address institutional, behavioural, and device-related factors that contribute to the occurrence of needlestick injuries. She finds that, despite significant progress in policy, practice and products, needlestick injuries continue to be a serious hazard, exposing health care workers to deadly viruses and other blood borne pathogens.

Chapter 10, by Carole Moderate, considers the ageing National Health Service workforce and ponders whether this is significant risk to the NHS and, therefore, to the nation. Apparently (and this may come as a surprise to some) an 'older worker' is anyone aged over 50! As nurses represent the majority of the NHS workforce, the chapter naturally focuses on that occupational group. The risk of a nurse staffing crisis is looming as a predicted 15% of the workforce is due to retire in the next ten years. Carole finds, however, that lack of empirical research into the effects and risks of the older workforce could be a risk in itself, as without this information we do not know how to address the problem. And if we don't address the ageing workforce issue, then we cannot, she says, ensure the continued safety of our patients, staff and the NHS as a whole.

Chapters 11 and 12 explore the topical issue of violence and aggression towards healthcare staff. Healthcare care organisations in the UK have a legal and ethical duty to protect staff from harm from foreseeable risk of violence. Carole Moderate and Becky Monaghan take different perspectives on violence and aggression towards healthcare, and particularly NHS staff, which is perceived by the general public to be increasing. Or is it a case that there is greater awareness amongst staff and healthcare organisations around violence and aggression, and reporting is better than it used to be? These chapters set out to explore this question and explain the reasoning behind why patients become aggressive or violent. Risk management strategies to deal with the problems of violence and aggression are also covered.

The staff theme continues in chapter 13, which looks at fire safety and the training of staff in fire prevention and management in healthcare premises. In this chapter, Jayne Hartley takes an empirical and pragmatic view of fire safety training in the NHS. In doing so, she considers whether the training provided for healthcare staff in the prevention and management of fire requires restructuring, and whether the current format for fire prevention and management training is research based and educationally sound. She concludes that such training leaves a lot to be desired and suggests a number of practical ways in which fire training might be made more meaningful and productive in the future.

Last, but by no means least, chapter 14, considers the contemporary but, to many in the NHS, less obvious risks (and opportunities) associated with disposal of surplus NHS buildings. The NHS estate is the biggest property portfolio in Europe, with many inherent risks and opportunities. In recent years the NHS has seen significant sales of surplus estate. Such sales, if handled properly, can realise significant benefits. However, if mismanaged, the risks of getting it wrong can result in missed opportunities for patients and staff as well as loss of potential income.

Concluding comments

In the rapidly changing world of the NHS, some aspects of the contents of this volume may, by the time it is published, be more contemporary than others. However, most, if not all of the issues were contemporary ten years ago, and some, if not most, might still be contemporary ten years hence. It remains to be seen just how the government and individual NHS organisations truly perform in managing contemporary healthcare risks.

On certain issues it may take many years to achieve the benefits of risk reduction. In the crucially important area of patient safety, for example, at the time of writing the National Patient Safety Agency (NPSA) has come in for some damning criticism from the Public Accounts Committee. According to the Committee's Chairman, the NPSA has exhibited "dysfunctional performance" in its failure over several years to establish a credible National Reporting and Learning System to learn from patient safety incidents and identify and promulgate solutions across the NHS. Not only has substantial public money been wasted by the Agency in pursuing an expensive system of its own making when other proven systems were available for use at far less cost and 'time to market', but, more importantly, many thousands of patients have suffered unnecessary death and other harm, some of which may have been prevented with solutions emanating from the national system. It remains to be seen what contemporary risks have been created, as opposed to mitigated, by the activities of the NPSA during its first five years of operation.

In the future volumes, we hope to bring to readers of healthcare risk material further published work around patient safety and many other issues in contemporary healthcare risk management. Over time we aspire to continuously improve the quality of risk information that is available to healthcare professionals, managers and board members. We would be grateful for you help. If you have any suggestions as to how we might make future improvements, no matter how small, or thought on emerging risk areas, then please contact either or both of us at the e-mail addresses given below.

One final comment. With the explosion of information available on the Internet, many references in published materials now relate to a web link and web links have an unfortunate habit of changing! All web links in this publication were checked and found working as at December 2007. Should you come across any instance where the link is no longer working, we would be grateful if you could inform us. Thank you.

Stuart Emslie - S.Emslie@lboro.ac.uk
Charles Hancock - C.P.Hancock@lboro.ac.uk
Loughborough University
January 2008

1

Consideration of the essential interacting elements of a healthcare risk management system

JANE RIPPON

Introduction

This chapter aims to identify the elements of a risk management system and discuss the impact that the interaction between these elements has on the risk management system within a real National Health Service (NHS) organisation. The organisation, which shall remain anonymous, but which I shall refer to as *Dingley Dell NHS Trust*, is a 650 bed rural district general hospital (DGH) providing acute hospital services, including Accident and Emergency Services, in south east England. The organisation has a dedicated 'risk management department', which forms part of a larger 'clinical governance support unit' (Figure 1.1).

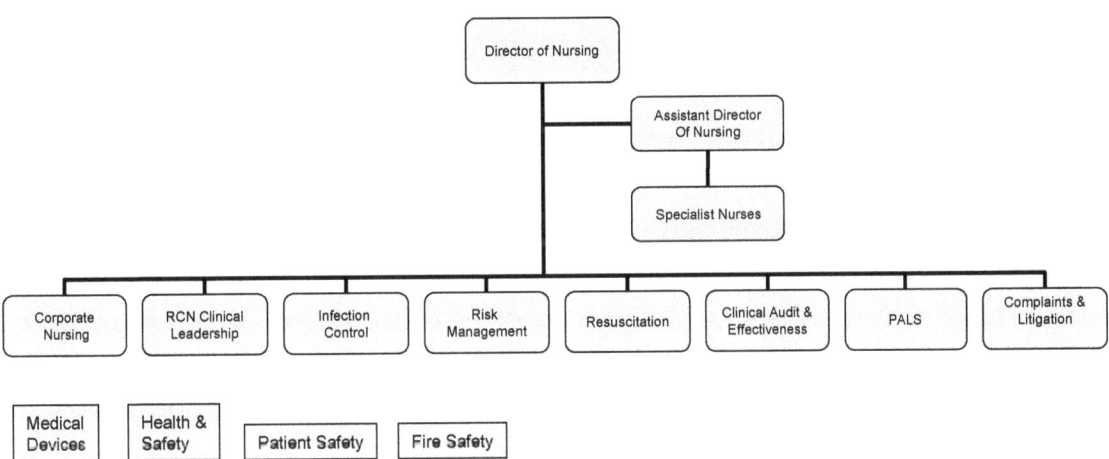

Figure 1.1. Dingley Dell NHS Trust Clinical Governance Support Unit Structure

An effective risk management system within any healthcare setting is reliant upon several interacting elements. The way in which these elements interact will, in many ways, determine the effectiveness of the risk management system (Office of Government Commerce, 2001).

The elements of a risk management system explored in this chapter are those identified by the Australian/New Zealand risk management standard 4360:2004 (Standards Australia and Standards New Zealand, 2004). Many of the local risk management systems in the NHS have been based on the risk management process described in this standard, and I shall look at each of the elements and how they need to interact if a robust risk management system is to be in place in an NHS healthcare organisation. In addition, the mechanisms that have been put in place in the NHS in England, both nationally and locally, to improve the effectiveness of each of the elements are explored. And the challenges experienced at Dingley Dell NHS Trust in trying to ensure that the elements combine and successfully interact to produce a robust risk management system are examined. Finally, consideration is given as to whether the approach taken at Dingley Dell has been successful.

Systems, risk and risk management

Checkland (1981) defines a *system* as a set of elements connected together which form a whole, thereby possessing properties of the whole rather than of its component parts. Senge (1990) went on to say that activity within a system is the result of the influence of one element on another, and the feedback can be either positive (amplifying) or negative (balancing) in nature. He also said that systems are not chains of linear cause and effect relationships, but complex networks of interrelationships. It is this argument that will be developed further in this chapter through assessing the risk management system that existed at Dingley Dell NHS Trust prior to April 2005.

The Australian/New Zealand risk management standard (Standards Australia and Standards New Zealand, 2004) describes risk as "the chance of something happening that will have an impact on objectives". It goes on to suggest that risk management can be defined as "the culture, processes and structures that are directed towards the effective management of potential opportunities and adverse effects."

Roberts (2002), takes the view that risk arises out of uncertainty from either an internal or external source and is the result of taking or not taking a particular course of action, which subsequently leads to the possibility of injury, delay, physical harm or financial loss or gain. In healthcare, there will always be risk attached to anything we choose to do. The choice, therefore, is between the actions we decide to take based on the level of risk we are prepared to accept either individually or as an organisation. The task of risk management in an organisational context is to limit the organisation's exposure to risk by taking action on the probability, or likelihood of the risk occurring, its likely impact, or consequences, or both.

The risk management process

With reference to Figure 1.2, the elements of the risk management process described in the Australian/New Zealand risk management standard (Standards Australia and Standards New Zealand, 2004) can be listed as follows:

- establishing the context
- risk identification
- risk analysis
- risk evaluation
- risk treatment
- monitoring and reviewing risks; and
- communication and consultation

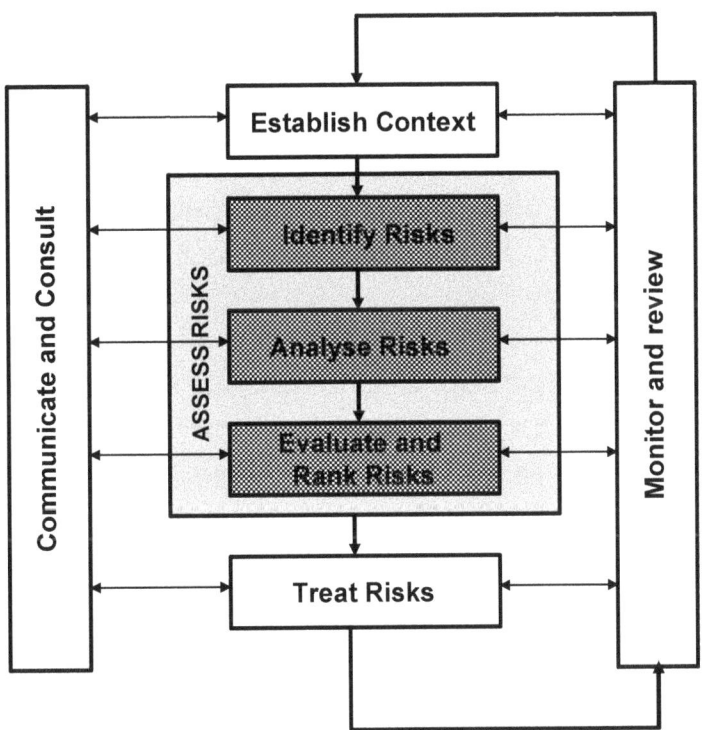

Figure 1.2 – AS/NZS 4360:2004 Risk Management Process

Each of these elements is now considered and, in the context of Dingley Dell NHS Trust, Table 1.1 identifies any products produced within that element, the stakeholders involved, the expected and actual outcome and the overall affect on the success of the risk management process within the hospital.

Establishing the context

The first step in the risk management process is to establish the context by, for example, considering the organisation's goals, objectives, values, policies and strategies. A complete understanding of the context within which a successful risk management system is to be implemented is crucial. Within this element, the current structure of the risk management department, the policies and procedures that exist within the organisation and the changes that have already been made within the organisation following a report by the then Commission for Health Improvement (CHI) in 2002 are explored [*Editor's note* – although, to preserve the anonymity of the Trust, the CHI report is not explicitly referenced in this chapter].

Risk identification, analysis and evaluation (i.e. 'risk assessment')

Hopkinson (2001) states that "the objective of the risk identification element is to ensure that all significant risks are listed so that they can be analysed and evaluated." This is a short step but a critical one to the process, as risks that are not identified will not be managed. Risk identification needs to include all key stakeholders, with all staff playing an active role in the process.

There are many methods of risk identification available, but the process needs to be simple if staff are to participate. Once risks are identified, action is taken to either prevent or control the risk. Failure to identify all risks could prevent the Trust from meeting its objectives; therefore risk identification should be a high priority within any NHS organisation.

Risk assessment is undertaken in many ways, through the incident reporting system, risk assessment process, Health and Safety Executive visits, Trust health and safety audit inspections, Clinical Negligence Scheme for Trusts (CNST) requirements and, formerly, through the Department of Health's Controls Assurance process (Department of Health, 1999).

The data collected from the risk identification phase has to be analysed to ensure that decisions can be made about prioritising and treating the risks. The Dingley Dell NHS Trust's system for risk analysis is a five by five risk scoring matrix, as used widely in the NHS, analysing the risks in terms of likelihood and consequence (see Figures 1.3 and 1.4). The analysis and evaluation process together enable the Trust to produce the information required to identify and select the actions that need to be taken in terms of risk treatment. It assists the organisation in separating minor, moderate and major risks.

Having analysed the risks, evaluating and prioritising the risks for action is usually straightforward. The risks are identified as being high, moderate, or low, and prioritisation is generally undertaken by senior management within the organisation, or by the risk management department. The prioritisation process helps the Trust to decide whether particular risks are acceptable or not, taking into account the controls already in place, and the financial consequences of managing the risk or leaving it untreated.

Risk treatment

The purpose of risk treatment is to determine what will be done to mitigate the risk, and who will be responsible for the treatment action(s). Risks are more likely to be acted upon if responsibility is allocated to an individual. Risk treatment options are evaluated in terms of feasibility, cost and benefits with the aim of choosing the most appropriate and practical way of reducing risk to a tolerable level. Risk action plans may seek to reduce the likelihood of occurrence, minimise the consequences, transfer or share the risk, or retain the risk.

Monitoring and reviewing risks

Continuous monitoring and review of risks ensures that new risks are detected and managed, action plans are implemented and managers and stakeholders kept informed. The availability of regular information on risks can assist in identifying trends like trouble spots or other changes that have arisen. It is essential that this information is accurate, complete and based on the most recently available data. Ongoing review is also required to ensure that risk treatment plans remain relevant. In reality, factors that impact on risk assessments are ever changing and it is therefore important for this process to be ongoing to ensure the risk management system remains effective.

Communication and consultation

At every stage of the risk management process, it is important that the risk management team communicate and consult with all the stakeholders in the process, both internal and external. All decisions should be made through consultation and should be effectively communicated to all stakeholders. This is essential to the success of the risk management system as it clarifies to each party the responsibility for risk, clarifies the nature and complexity of specific risks and increases overall confidence in the risk management process.

Practical risk management at Dingley Dell NHS Trust

In 2002, a CHI report identified that the Trust had a split risk management function with clinical and non clinical risk being dealt with through separate departments and separate reporting systems. As a result of the CHI report, the risk management department was set up and the risk assessment and incident reporting systems were unified. Trust staff now have one form for all untoward incidents and one form for all risks identified (risk assessments). Directorates have produced their own clinical governance frameworks within which they have a risk management committee. The Trust is striving towards a risk management culture within the organisation and policies and procedures have been developed to support the process. However, the mechanism for ensuring staff are aware of these policies is not well developed.

All new policies and procedures are published on the Trust intranet and hard copies are forwarded to all wards and departments as appropriate. But who ensures staff are aware of these policies? Do they read them and do they know where to find them? Some 60% of wards in the hospital do not have access to the hospital intranet, and therefore rely on hard copies for their information.

In the medical directorate, lack of knowledge of new policies and procedures was identified as a risk and a new system was implemented on all medical wards. Each ward has a purple folder and all new policies and procedures are placed in the purple folders until such time as staff have read them and signed to say they have done so. The system has been in place for four months and a 'spot check' audit was undertaken by the patient safety manager. It was found that on four of the seven medical wards checked, 80% of staff had not looked at the purple folder during the last four months.

From the point of view of staff working within the organisation, risk identification is their major contribution to the risk management process. Incident reporting forms part of the Trust induction training programme for all staff, and is also part of the annual mandatory risk management update training. It is given a high priority by the organisation and time has been spent producing training programmes for staff to ensure that they are fully conversant with the incident reporting system. However, in some areas reporting is more successful than in others. Similarly, in some groups of staff, reporting incidents and undertaking risk assessments is more developed than in others. In 2003/04, 80% of incidents and risk assessments (including clinical and non clinical risk and patient falls) were completed by nursing staff, 8% by medical staff and 12% by other staff in the organisation.

The Trust uses a computerised information system to collate the information provided in the incident reports and risk assessments. This enables the risk management department to undertake trend analysis and produce monthly reports to each directorate on all incidents and risk assessments received. However, the system was implemented in haste prior to a CNST inspection and, as a consequence, very little training for staff accompanied the implementation. This meant that many of the forms that were received in the risk management department were incomplete and coded incorrectly. This issue is now being resolved through the ongoing risk management training programmes that exist, but will take some time to filter down to all staff in the organisation.

Designated staff in each area of the hospital have undergone risk assessment training and as part of this training have been educated on how to use the Trust risk scoring matrix. However, the risk assessment training sessions are only updated annually and, if staff leave their employment with the hospital, there are times when wards or departments have no formally trained risk assessors in their areas. This means that when incident reports and risk assessments arrive in the risk management department, risk scores are either absent, or have been incorrectly completed. In these cases, the risk management department have to score the risks on behalf of that department. There are policies and procedures in place relating to untoward incident reporting and risk assessment and the risk scoring matrix forms part of both of these policies. However, as already mentioned, staff have often not taken the appropriate steps to familiarise themselves with the contents of the policies.

Table 1.1a – Dingley Dell NHS Trust: Consideration of elements of risk management system in relation to key issues

	Establishing the context	Risk identification	Risk analysis	Risk evaluation
Products produced	Policies and Procedures Training Programmes Risk Management Strategy Organisational Objectives	Incident Reports Risk Assessments Serious Untoward Incident Reports Adverse Staffing Forms Reporting of Injuries, Diseases and Dangerous Occurrences Regulations (RIDDOR) Reports	Risk score for each incident and risk assessment Risk Evaluation	Trust Risk Register Directorate Risk Register Serious Untoward Incident Reports Root Cause Analysis investigations
Stakeholders involved	All Trust staff Trust Board Strategic Health Authority Patients Visitors and the general public Contractors Locum and agency staff CNST National Patient Safety Agency Primary Care Trust Healthcare Commission	All Staff Contractors working on the premises Agency and Locum Staff Patients Visitors General Public Trust Board Strategic Health Authority National Patient Safety Agency Health and Safety Executive	Directorate management teams Risk Management department Trust Board Risk Management committee	All Staff Patients Visitors General Public Trust Board Strategic Health Authority National Patient Safety Agency
Expected outcome	Organisational culture supportive of risk management. Risk management embedded into the infrastructure of the organisation with all staff aware of their contribution to the process.	Identification of all major, minor and moderate risks Identification of potential health and safety issues Identification of near miss events	Each risk analysed in terms of likelihood and consequence to give it a risk score. This risk score then produces a low, medium or high risk evaluation which the Trust can use to manage the risks appropriately	Risks appropriately categorised into high risk (which form part of the Trust risk register), medium risk (which form part of the directorate risk register) and low risk (which form part of the risk register for the individual department assessing the risk, but will be reviewed periodically by the directorate)
Actual outcome	Risk management is becoming more embedded into the organisation, but there are still pockets of staff who see the risk management department as responsible for risk, not seeing themselves as stakeholders in the process.	4,000 clinical risk issues identified annually 1,500 non clinical risk issues identified annually Some areas better at reporting than others Some areas better at producing risk assessments than others	Each risk is analysed and given a risk score. However, senior managers will often over-inflate a risk score to push it higher on the Trust agenda and similarly will often under-score risks that they feel are "bound to occur" – this is particularly true in the surgical directorate with risks relating to surgical procedures. The system is quite subjective and often risk scores are recalculated by the risk management department.	The Trust risk register contains all high risk evaluations throughout the organisation. These are monitored by the risk management committee on a bi-monthly basis through specialty reviews. Each directorate is expected to provide evidence of action plans for each risk on the risk register. Each directorate also has a risk register containing moderate risks; and action plans for these are monitored at directorate clinical governance committee meetings twice yearly. Ward and departmental risk registers are much patchier. In some areas of the Trust they are fully operational and working well to manage low level risks in these areas, in other departments the risk register exists but as a purely tick box exercise
Overall effect on the success of the risk management process	Staff, particularly medical staff, often view risk management as a bureaucratic paper exercise that takes them away from their proper jobs. Although risk management is an element of all senior staff job descriptions, no additional time has been allocated for this important function. Therefore managers are more often reactive to risk when incidents occur, rather than being proactive about managing the sources of risk within their areas.	Identification of risk is crucial to the success of the risk management process; the other elements of risk management are dependent upon risk identification to enable them to function.	For the process to continue from the risk identification state, some form of risk analysis is necessary to enable the Trust to put risks into a priority listing.	Risk evaluation relies on the elements of risk identification and risk analysis, but also relies on the staff within the organisation to comply with Trust policies and procedures relating to monitoring of the risk registers once produced. This element tends to be more overlooked as the Trust can only monitor the high risk areas identified as part of the Trust risk register.

Table 1.1b – Dingley Dell NHS Trust: Consideration of elements of risk management system in relation to key issues cont.

	Risk treatment	Monitoring and review	Communication and consultation
Products produced	Action plans which will include risk treatment options of: 　Risk avoidance 　Risk acceptance 　Risk reduction 　Risk transfer	Monthly reports to departments Monthly reports to directorates Reports to Trust committees Reports to the strategic health authority Reports to the health and safety executive Audit reports Ad hoc reports as requested by members of the organisation Monthly reports to National Patient Safety Agency Root cause analysis reports and investigation reports	Monthly update reports for all wards and departments Risk management newsletter Quarterly reports to Trust Board and Risk Management Committee Quarterly reports to directorate clinical governance meetings. Training sessions and risk management workshops
Stakeholders involved	Trust Board Risk Management Committee Directorate Management Teams Department and Ward management teams	All Staff Trust Board Strategic Health Authority National Patient Safety Agency Health and Safety Executive Primary Care Trust	All Staff Trust Board Directorate Management Teams
Expected outcome	Risk scores reduced by treatment plans implemented	As risks change, new measures are introduced to control these risks. Accurate summarised information based on latest available data is provided to senior Trust personnel to ensure management and treatment plans remain relevant.	Staff feel involved in and understand the risk management process. They understand their own roles within the risk management process and can see the benefit of completing incident forms and risk assessments.
Actual outcome	Financial and resource considerations often mean that risks are not treated at source, but untoward incidents that occur as a result of an identified risk may push the risk up the priority list for action to be taken. However, the current financial pressure means that any risks requiring additional funding may remain untreated on the risk register and be accepted by the Trust board.	Monthly reports are produced by the risk management department. Action plans are produced for all incidents reported and risk assessments but are monitored by the directorates themselves or by the Trust board and not by the risk management department.	Many front line staff feel they get little or no feedback, despite the fact that monthly reports are produced for managers to cascade down to ward and department staff.
Overall effect on the success of the risk management process	This element of risk management is currently managed outside the risk management department, which makes it very difficult to assess its effect on the risk management process. However, the fact that many identified risks remain untreated does cause a loss of faith in the risk management process within the hospital and has the potential to undermine its success.	This element interacts with and is highly dependent upon the communication element which is weak within the organisation.	This is the least successful of all of the elements of risk management and is currently the biggest barrier faced by the risk management team.

All risks identified are graded as low, medium or high risk events (Figures 1.3 and 1.4). However, despite the risk evaluation guidance shown in Figure 1.4, the risk scoring is very subjective and often wards or departments will allocate a high risk score to a problem that the risk management department feels is a much lower risk. This is due to the fact that risks are generally identified reactively rather than proactively as previously mentioned. Therefore, once an untoward incident has occurred, the atmosphere is quite emotive and therefore a higher risk score is placed on the untoward event. In the cold light of day, however, it becomes apparent that either the likelihood or consequence score has in fact been over inflated somewhat and the actual score is much lower, placing the risk in a lower risk band.

Consequence	Likelihood				
	Rare 1	Unlikely 2	Possible 3	Likely 4	Almost Certain 5
Catastrophic - 5	5	10	15	20	25
Severe - 4	4	8	12	16	20
Moderate - 3	3	6	9	12	15
Slight - 2	2	4	6	8	10
Low - 1	1	2	3	4	5

Green — Low risk events. Investigated at discretion of Directorate Manager/Head of Dept.

Amber — Must be investigated by the Head of Dept. An action plan must be prepared and forwarded to Risk Management

Red — Must be communicated to the Risk Management Department as soon as possible

Figure 1.3 Dingley Dell NHS Trust Risk Scoring Matrix

The clinical risk management section of the CHI report stated that "the Trust board needs to be more responsive and receptive as there is discontent amongst some clinicians that identified risks are not dealt with effectively and that action is only taken when a crisis is reached." The report went on to mention that it was not clear how actions are implemented Trust wide or what happens if investment is not available. CHI also said that the organisation needed a formal methodology for prioritising corrective action and a transparent process so that staff could be better informed and aware of the rationale behind decision making.

Because the Trust is reactive to risk rather than proactive, it means that often risk treatment is limited to fixing the causes of incidents rather than systematically and proactively identifying and treating risks at source, i.e. before the risk materialises as an incident. Due to financial constraints, solutions are often required to have a cost saving or neutral cost attached to them, and often risk management solutions identified will initially cost money to eventually save money or time or resources. Failure of the Trust Board to buy into this way of thinking causes great frustration with the staff working within the risk management department and also those out on the wards.

Measures of consequence

Level	Descriptor	Description
5	Catastrophic	Death of a person
4	Major	Significant injury, harm or permanent disability
3	Moderate	Excessive injuries requiring corrective surgery or treatment and extended length of stay for the patient. Some loss of reputation.
2	Minor	First aid treatment only required. No lasting harm to injured party and no extended length of stay.
1	Insignificant	No injury caused.

Measures of likelihood

Level	Descriptor	Description
5	Almost Certain	Happens on a daily basis or 1:10 cases
4	Likely	Happens at least once a week or 1:100 cases
3	Possible	Happens at least once a year or 1:1000 cases
2	Unlikely	Happens every 2-5 years or 1:10000 cases
1	Rare	Only happens once every 5 years or 1:100000 cases

Figure 1.4 Dingley Dell risk evaluation information

Interestingly, the lack of action by the Trust in relation to wider risk management has not led to a decrease in the number of incidents reported, apart from within the medical staff group. From 2001 to 2004 incident reports have steadily increased by about 11% a year.

The risk management department produces monthly reports and trend analysis for all directorates and for the clinical risk, medical devices, health and safety and risk management committees and the Trust Board. However, the usefulness of these reports is often debated as they contain so much information it is difficult for them to be properly monitored during a two hour committee meeting. In practice what tends to happen is that the directorates are expected to monitor their own reports and the Trust committees will monitor only the risks designated by the risk evaluation as high. This means that trends within moderate or low risk areas are often missed, as directorates tend to look at individual risks rather than trends within risks.

Monthly feedback reports are prepared for each directorate for all incidents that occur during that month. These reports should then be cascaded down to 'front line' staff through monthly meetings, ward meetings etc. There is no evidence that this happens, indeed many staff still feel they receive little or no feedback from incidents reported and it could therefore be assumed that incident reports are not discussed at these meetings. On questioning staff, however, I found that incidents are discussed regularly at meetings, and the "no feedback" the staff refer to simply means that nothing has been done about the incident in question, i.e. no risk treatment action has been taken by the risk management department. This is particularly true in relation to incidents reported by the senior medical staff within the hospital, who feel that the completion of an incident form means the risk has been transferred to the risk management department who will deal with that risk, and do not.

Communication within the Trust is hampered by the fact that not all staff have access to the Trust intranet system. Many of the wards still do not have computer links, which means that information has to be cascaded down through the management structure, rather than giving staff the opportunity to use the intranet site to keep up to date.

The Trust is very much divided into silos – each directorate works as an entity within its own right and there is very little communication between directorates or sharing of learning following the identification of risk issues. This has, to some extent been addressed by the risk management newsletter but, following the introduction of the Freedom of Information Act (FoI), this may be a less useful route of communication as all information contained in the newsletter will be in the public domain and therefore may be 'watered down'.

Concluding comments

Establishing the context is the linchpin of all of the elements of risk management. If the risk management is not placed at the heart of the organisation's objectives, the elements of risk management will not interact sufficiently to ensure a robust risk management system within the organisation.

The appropriate policies and procedures are in place but these are very much centrally driven and are not owned by the directorates. Indeed, staff within the organisation often do not feel it important to acquaint themselves with new policies and procedures as they are published. And the fact that many wards and departments do not have intranet access also impedes the ability of staff to access policies and procedures.

If staff do not receive the appropriate training, then it is difficult for them to fully involve themselves in the risk management process. For this reason, the Trust has invested time in developing a mandatory risk management training programme which is undertaken by all hospital staff on an annual basis. In order to capture the medical staff, this mandatory training is given during the monthly clinical governance afternoons (formerly the audit afternoons). These sessions remain protected time for medical staff with clinics and theatre lists cancelled and therefore good attendance is achieved.

Risk identification links and interacts with risk evaluation, risk monitoring and risk communication. If staff do not feel the feedback mechanisms are working appropriately, they will stop completing incident forms and risk assessments.
Lots of time is spent on these three elements and some initial treatment of risks is undertaken by wards and departments. Some risks are reported expecting the risk management department to take action on their behalf. Root cause analysis is not undertaken in all wards or areas and is only done by the risk management department for serious untoward incidents. Often only superficial action is taken following an incident. The current culture is to be reactive to the incident occurring rather than proactively trying to stop the incident from recurring and treating the risk at source. There is no budget within the risk management department and therefore no quick fix solutions available for minor problems.

The interaction between risk evaluation and the other elements of risk management within the Trust is well developed for high level risks. It is less well developed however for moderate or low risk events as this is dependent upon wards, departments and directorates producing action plans, monitoring those action plans and communicating and consulting with their staff.

The culture within the organisation is very much a 'fire fighting' culture. With little or no spare resources, only limited proactive risk treatment at source is undertaken, and this tends to be in areas related to health and safety issues, rather than issues of patient safety or clinical risk. The managers within the organisation still see risk management as something extra they have to do, rather than being an integral part of their role. There is no additional time allocated for them to undertake the tasks required within the risk management element of their job descriptions and therefore only the bare minimum required to satisfy Trust and national requirements is undertaken.

Monitoring and review of risk is done by the various groups within the Trust. These activities appear to take place in silos, and, although the risk management department is nominally responsible for monitoring all risks, this is not always the case, due to the many and varied committees that review action plans.

Communication links all the other elements. There are several problems with communication within the hospital, not least of which is the fact that not all areas of the Trust are able to access the hospital intranet. Risk issues are discussed at many forums within the organisation but there is still a perception for some staff that risk is not their responsibility and that is why we have a risk management department. This perception is changing but it is a slow process, particularly as much of the workforce is quite static in nature and have worked in the hospital for some years. This makes it much more difficult to implement changes and adapt the way staff view the whole process. Communication largely relies on information being cascaded down to 'front line' staff by the management team in each directorate. The risk management department also produces a bi-monthly newsletter called 'Risk Matters' which is published on the Trust intranet. As a result of discussions with ward staff, the risk management newsletter is now also produced in hard copy format and placed on all ward notice boards for staff to read. Staff are able to submit copy for the newsletter relating to any risk issues they feel need to be highlighted across the Trust. The newsletter is still in its infancy but very favourable feedback has been received so far on the three issues produced to date.

The different elements of the risk management process do not have equal weight or importance within the Trust, but they are all crucial to the success of the risk management systems within the organisation. Much of the work of the risk management department is currently focussed on risk identification and analysis and very little time is spent on risk treatment and risk monitoring. This means that the Trust is very reactive when it comes to dealing with risk issues. There is very little pro-active risk management taking place in the organisation and risks are often left untreated, either for financial reasons or because of the lack of understanding by the Board of the benefit of treating risks at source.

Hopkinson and Hopkinson (2001) state that a good risk management culture is essential as without this there will be either ignorance about the importance of risk identification or worse, unwillingness to divulge new information. The Hospital does not have a very fluid work force. Many staff have lived and worked in Dingley Dell for all or most of their working lives, indeed many of the nursing staff trained in Dingley Dell. This makes it much more difficult to change the culture as some staff are set in their ways and can often be regarded as 'retired in post'.

From an organisational point of view all of the interacting elements of risk management are now within one directorate, the clinical governance support unit. However, complaints and litigation, clinical audit and effectiveness and infection control currently sit outside the risk management structure and therefore do not contribute to the risk management process.

Most of the work undertaken by the risk management department has been driven by the past requirements of Controls Assurance and CHI and, more recently, by the requirements of CNST and the Healthcare Commission. Risk management has not been driven by the Trust's belief in the principles of managing risk. All the elements of risk management are present within the organisation, but do not interact successfully, and therefore the system has major flaws. However, the Trust has made significant progress towards an integrated system of risk management and continues to move in this direction.

References

Checkland, P (1981). *Systems Thinking, Systems Practice.* New York: Wiley.

Department of Health (1999). HSC 1999/123, Governance in the new NHS. Controls Assurance Statements 1999/2000: Risk Management and Organisational Controls. Department of Health, London.

Hopkinson M and Hopkinson J (2001). Adapting techniques to manage risk, *Health Care Risk Report* 7(2):19-20.

Office of Government Commerce (2004). *Successful Delivery Toolkit Version 4.5.2.* HMSO, London.

Roberts G (2002). *Risk Management in Healthcare.* 2nd Edition. Witherbys Publishers, London.

Senge, P. 1990. *The Fifth Discipline: the art and practice of the learning organisation.* London: Doubleday/Century Business.

Standards Australia and Standards New Zealand (2004). *Risk Management.* AS/NZS 4360:2004. www.riskmanagement.com.au

2

A review of the extent to which the Trust Risk Management Strategy is functioning within the emergency services division of an NHS Trust

CAROLE MODERATE

Introduction

Good risk management awareness and practice at all levels is a critical success factor for any organisation. In healthcare it can mean the difference between success and failure, not only in terms of an individual patient clinical outcome, but also of the organisation as a whole (Roberts, 2002).

Effective risk management is dependent on:

> (a) establishing a corporate and systematic process for evaluating and addressing the impact of risk in a cost effective way; and
> (b) having appropriately trained staff with the skills and knowledge to identify and assess the potential for risk to occur and manage it.

A risk management strategy provides a framework for the development and implementation of a rigorous risk management process throughout an organisation, and is a sign that the organisation is committed to continuous quality improvement (Roberts, 2002). If the strategy is to be a success, people must be engaged with both hearts and minds. The consequence of failing to engage people at all levels in the planning stage of the strategy can result in lack of ownership and a failure of the organisation to achieve its strategic goals (Semple-Piggott, 2000).

In this chapter, I report on a 'snapshot' review of the functioning of a Trust risk management strategy in the emergency services division of a medium-sized NHS Trust. The Trust comprises a 400-bed district general hospital that provides acute health services to people in an area in the south east of England.

Methodology

Using the overall statements of intent contained within the Trust Risk Management Strategy as a basis for setting criteria against which to measure, two staff questionnaires were designed in order to assess how the nursing staff at all levels perceived they were applying the risk management tools described in the Risk Management Strategy within their clinical area. One questionnaire was aimed at A–E grade nurses, focusing on issues such as incident reporting, risk assessment and general risk awareness. The other questionnaire was aimed at F and G grade nurses, in the roles of ward managers and deputies, to assess whether a more strategic approach to risk management was applied at this level. The second questionnaire focussed on the issues of sharing the feedback and ensuring lessons learnt are communicated across the whole team. In both questionnaires, nursing staff were asked whether they felt the organisation had provided them with the key skills and training to manage risks effectively.

The clinical areas under review in this chapter are three acute medical wards within the Emergency Services Division, which is the largest division in the Trust. The three wards are similar to one another in terms of numbers of beds, environmental design, staffing establishment and patient type. As such, for the purpose of this chapter, the three wards are classed as one clinical area.

Forty-five of the first questionnaires were given out randomly to nurse grades A–E. The questionnaires were left in a staff room with no particular emphasis on who should complete them. A general explanation was provided and a time frame of two weeks given for completion. A total of six of the second questionnaire were given out individually to the F and G grade nurses who, at the time, had managerial responsibility for the clinical areas.

Results and Discussion

Out of 45 questionnaires given to the grades A–E, 25 were returned giving a 56% response rate. Four out of six surveys were returned from the F and G grades, giving a 67% return rate.

Responses from the survey were grouped using a model provided by Dale and Woods (2000) that describes an effective risk management strategy as having six key inter-relating components:

1. Organisational issues
2. Cultural issues
3. Clinical issues

4. Employee issues
5. Environmental issues
6. Incident reporting

As a separate exercise, not reported here, the model was used to translate the key findings from the survey into a management report.

Organisational issues

Dale and Woods (2000) argue that it is a fundamental requirement of a risk management strategy that risk management should be seen and embraced as a key area of line management responsibility. Managers at all levels should believe in this approach and take ownership for the proactive and reactive management of risk within their area of responsibility. Dale and Woods (2000) identify the following additional organisational attributes as facilitating an effective risk management strategy:

- An integrated, multidisciplinary risk management team, responsible for ensuring the strategy objectives are implemented
- Overall risk awareness of all levels of staff
- Risk identification, assessments, control and monitoring systems clearly communicated and understood

Understanding roles and responsibilities relating to risk management and an awareness of the strategic direction of an organisation are therefore key to its success.

The Trust Risk Management Strategy clearly states that the Chief Executive is responsible for risk management within the trust. However, only fifty per cent of the ward managers correctly identified the Chief Executive as being overall responsible for risk management and the other 50% didn't know who was responsible. Only eight per cent of A-E grade nurses identified the Chief Executive as being overall responsible for risk management. Forty four per cent of A-E grades thought responsibility for risk management lay with the Clinical Risk Manager.

All the ward managers were aware that the trust had a Risk Management Strategy, although the A-E grade responses were split (Figure 2.1).

Figure 2.1

Both groups were asked if there was a copy of the strategy in the clinical area; it is clear from Figure 2.2 that many staff were unable to say whether the strategy was available in the clinical area.

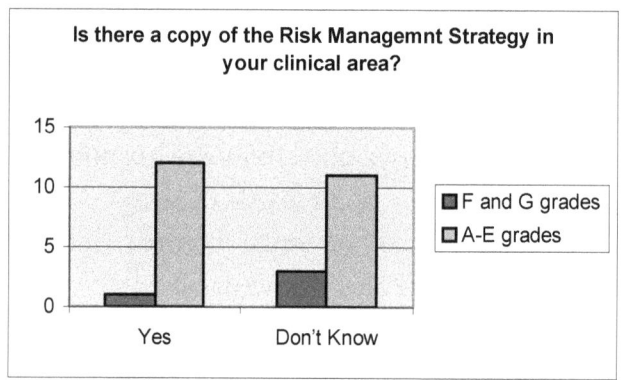

Figure 2.2

Both groups of staff were asked, "How important is risk management as part of your role." All of the managers responded "very important". By comparison, the A-E grade responses showed a lack of risk awareness, with seventy two per cent responding "not important" and four per cent responding "Not part of my role."

Cultural Issues

Risks within an organisation are reduced if an attitude of openness and honesty is adopted (Dale and Woods 2000). Harris (2000) also reviewed how the organisational culture facilitated a reporting and learning environment.

Both managers and A-E grades were asked if they agreed that incidents are investigated within a "fair blame culture." Figure 2.3 indicates a divergence in opinion between the two groups, with managers tending to agree and A-E grades tending to disagree – some strongly.

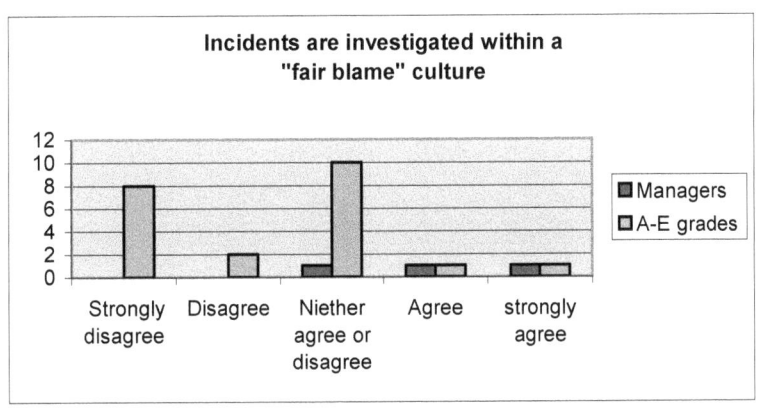

Figure 2.3

Both staff groups were asked how important they perceived *"patient"* and *"staff"* safety to be in the organisation. The managers unanimously agreed that patient and staff safety were equally 'very important' to the organisation. The A-E grades however, were more split in their opinions as can be seen in Figure 2.4.

	How important do you feel patient safety is in the organisation? (A-E grades)					
Not important	1	2	3	4	5	Very Important
No. of responses	0	1	1	4	18	

	How important do you feel staff safety is in the organisation? (A-E grades)					
Not important	1	2	3	4	5	Very Important
No. of responses	3	2	1	1	17	

Figure 2.4 – Patient safety vs. staff safety: responses from A-E grades

According to Dale and Woods (2000), poor communication is one of the highest risk factors in any organisation. The A-E grades were asked at what forum their line manager discussed with them the Trust Risk Management Strategy discussed with them. The majority responded that it had never been discussed (Figure 2.5). This compared with 50% of the managers, who stated they had never had the trust Risk Management Strategy discussed with them. This lack of communication represents a barrier to implementing the objectives in the strategy (Johnson and Scholes, 2002).

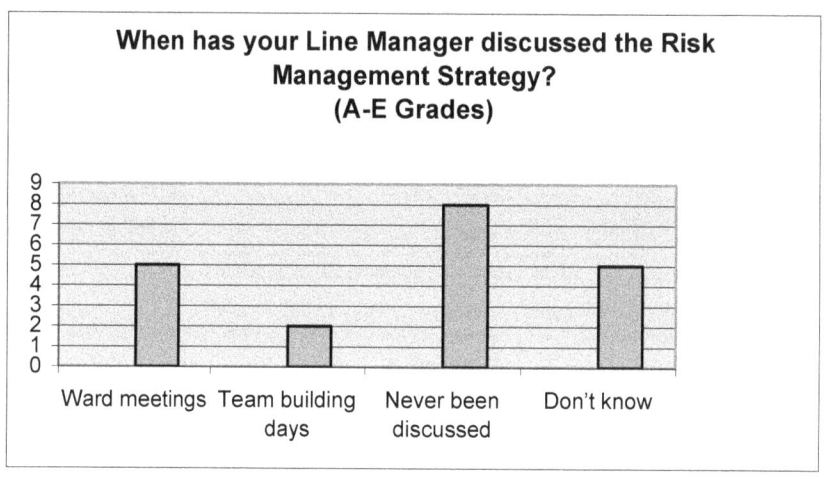

Figure 2.5

Clinical Issues

The managers were asked questions about risk assessments (Figure 2.6), and the results indicated that the process for proactive identification of risks was being followed, but the inconsistency of responses regarding the departmental risk registers demonstrated a lack of overall confidence and knowledge of the process. Dale and Woods (2000) state that identification and assessment processes are vital to the success of the organisation's risk management strategy. Harris (2000) also found in a comparative study that, although there was a high level of risk awareness, there was lack of clarity about individual arrangements and responsibilities and the active use of risk assessments as a local tool for managing risk.

How often do you complete risk assessments in your area? (F&G grades)	
Monthly	25%
Yearly	25%
Other	50%
How often do you review risk assessments in your area? (F&G grades)	
Yearly	75%
Never	25%
How many risk assessments are on your departmental risk register? (F&G grades)	
11-20	25%
Don't know	75%

Figure 2.6

The A-E grades were also asked if they had been involved in completing a risk assessment (Figure 2.7). Most had never been involved.

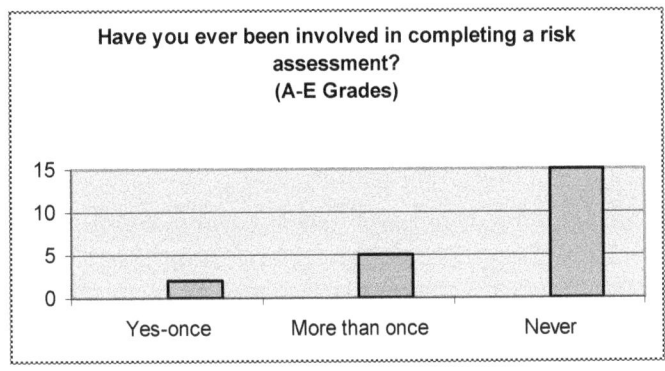

Figure 2.7

In addition, the majority of A-E grades asked were unaware their department had a local risk register (Figure 2.8).

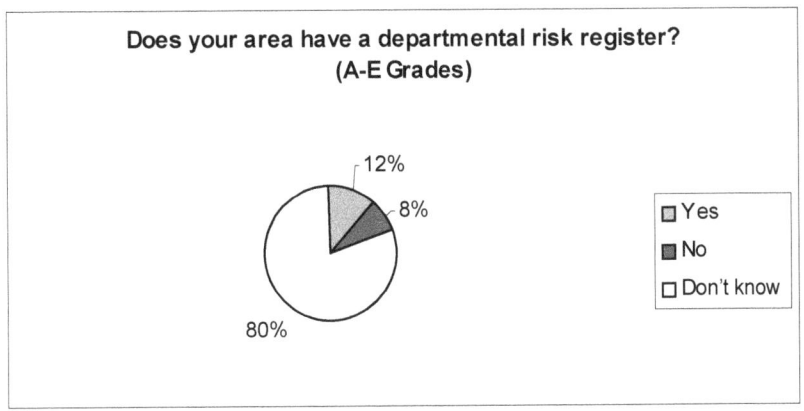

Figure 2.8

The Trust Risk Management Strategy states that risk assessments numerically graded greater than eight (>8) should be entered onto the Trust Risk Register. Emslie (2004) discusses the importance of making the Trust Board and Chief Executive aware of the significant risks an organisation faces and the controls in place to minimise the risk, as this information is vital to fulfil the requirements of the 'Statement on Internal Control' (Emslie, 2004).

Managers were asked what risk grades were required to be entered onto the Trust Risk Register (Figure 2.9). The fact that 50% of the managers were unaware that a risk assessment graded >8 should be entered onto the trust Risk Register brings into question the accuracy of the information regarding significant risks in the organisation that is regularly presented to the Trust

Board. The Board cannot be assured that all significant risks are being reported to it through the Trust Risk Register.

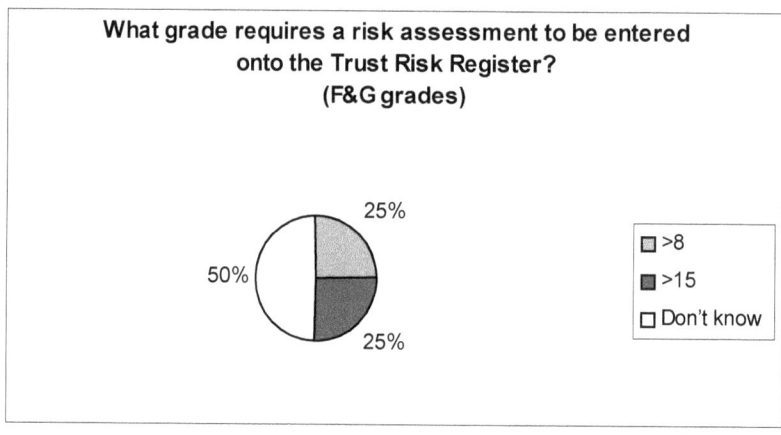

Figure 2.9

Employee Issues

Anderson (2004) states that people are at the "heart "of a strategy; individuals need to be given the resources, including training, to help a strategy succeed. Dale and Woods (2000) also stress the importance of staff development and education. Results from the survey indicate there are training needs across all levels of staff within the area (Figure 2.10).

When did you receive annual updates for the following training?								
	In last 6 months		In last 12 months		Over 12months		Never	
	Managers	Staff A-E	Managers	Staff A-E	Managers	Staff A-E	Managers	Staff A-E
Manual Handling	2	17	2	3	0	4	0	0
Fire	2	16	2	4	0	5	0	0
CPR	3	16	1	4	0	4	0	1
Infection Control	2	5	2	4	0	4	0	10
Blood products collection	1	13	1	6	3	5	0	1

Figure 2.10

Over half of the A-E grades stated they had never had training in risk assessment and a quarter had never been trained to complete an incident form. Fifty per cent of the managers had never had training to do a risk assessment. There was, however, a very positive response from all levels of staff when asked if they would like training (Figure 2.11).

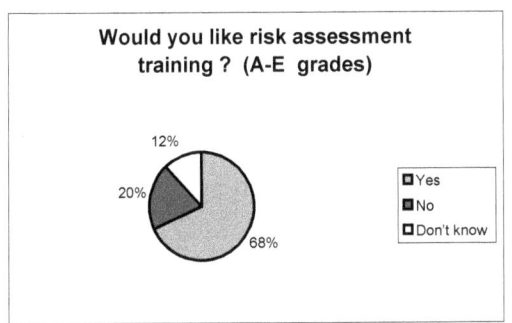 

Figure 2.11

Environmental Issues

Dale and Woods (2000) suggest that organisations can only achieve compliance with current Health and Safety legislation by adopting a well-managed strategic approach. The Trust Risk Management Strategy states that appropriate risk management policies should be in place and it is the responsibility of managers to ensure these are communicated. Dale and Woods (2000) imply that a significant cause of risk is due to staff being unaware of the organisational policies and procedures.

Thirty-two per cent of the A-E grades did not know where the policies relating to risk were kept in their clinical areas, and the results suggest a lack of awareness of the policies (Figure 2.12).

Which of the following Risk Management Policies have you seen A-E grades)			
Incident and Near Miss Policy	52%	Manual Handling Policy	76%
Risk Assessment Policy	40%	Fire Policy	80%
Health and Safety Policy	72%	Violence and Aggression Policy	20%

Figure 2.12

Incident reporting

Dale and Woods (2000) highlight that identifying risks retrospectively and analysing why things go wrong, in order to identify lessons learnt and action plan to prevent recurrence, is an important part of the risk management strategy. On the other hand, Lugon (2003) argues that although a good risk management strategy will encourage staff to report incidents, it does not guarantee that all incidents will actually be reported.

The A-E grades were asked if they had ever filled out an incident form. Although 52% stated they had completed an incident form more than once, 20% stated they had never completed an incident form. They were also asked if incident reporting is covered in the local induction process (Figure 2.13). The results suggest a lack of clarity regarding risk management support within the clinical area.

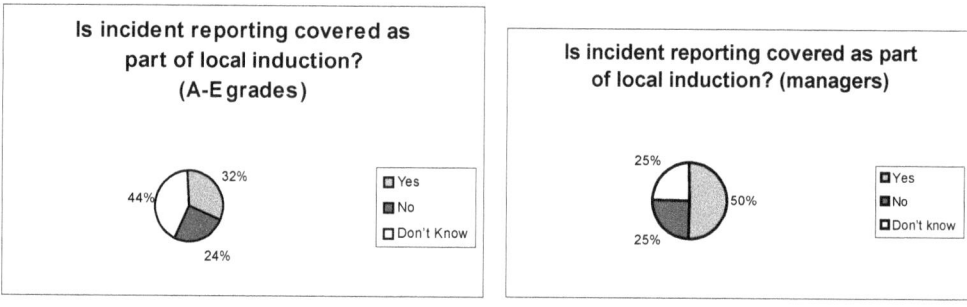

Figure 2.13

Dale and Woods (2000) suggest that the number of incident reports will increase if feedback is provided to the reporters. Lugon (2003) supports this argument, and suggests that the incident reporting process can only be successful if the team have ownership of the incidents and are empowered to reflect and learn from timely feedback.

Both A-E grades and managers were asked how often the team reviews incident report form and at what forum. The results are presented in Figures 2.14 and 2.15, respectively. It is clear from the results that there is a difference of opinion between the two staff groups about how often, and in what forum, incidents are reviewed and fed back within the team. In a separate response, all of the managers stated they occasionally discussed feedback from incidents with the staff

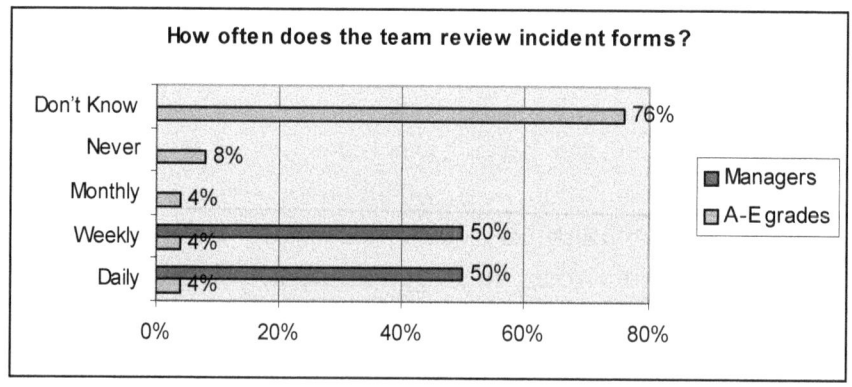

Figure 2.14

When does the team discuss/review incidents?		
	A-E grades	Managers
Ward meetings	20%	50%
Team brief	0	0
Handover times	12%	0
Don't discuss/review incidents as a team	20%	25%
Ward Communication book	0	25%
Ad hoc meetings for specific incidents	0	25%
Don't Know	52%	0

Figure 2.15

Conclusions

The review presented in this chapter has provided insights into the effectiveness of the Trust Risk Management Strategy through assessing staff perception of risk and risk management within a local clinical area. The focus of the review was on the overall risk management process identified in the strategy, rather than measuring against the specific annual objectives. This method of evaluation was chosen because the defined objectives relate to strategic functions rather than the application of risk management methods at a local level. Wilson (1997) suggests that the effectiveness of a risk management strategy can be gauged through the application of risk management tools. On the other hand, Anderson (2004) states that the true impact of a risk strategy is difficult to measure, and suggests quantitative benchmarks of a reduction in incidents, complaints or claims. A qualitative benchmark of assessing the staff's perception of risk issues in the organisation and measuring for an improvement in the safety culture may be a better test for assessing how well the strategy is functioning.

There is a need for a more strategic approach to implementing risk management processes at a local level. The evidence from this review indicates that perception of how risk issues are communicated within the team varies from one staff group to the other. This variation suggests ineffective communication and a lack of ownership. Despite the introduction by the Trust in the year preceding the review of a briefing paper, to be used as a tool to assist management communicate the strategy, only 65% of the A-E grades were aware that the trust even had a Risk Management Strategy.

Semple–Piggot (2000) suggests that the writing of a strategy is only the start of the process and suggests that regular reviews, monitoring or assessment should take place, with objectives being monitored against a constantly changing environment. Wilson (1997) also argues that, to ensure the impact of various risk management activities is measured accurately, the process for evaluation should be a multidisciplinary approach, engaging clinicians at all levels, the executive team and patients and public representation. Furthermore, Semple-Piggot (2000) goes on to suggest that a successful risk strategy is one with a shared vision. This argument fails when staff are either unaware of the vision or have no commitment to seeing it succeed. The Trust Board minutes for the present financial year (2004-05) have been reviewed, and no evidence was found that there was any formal review of the trust Risk Management Strategy carried out during this period. This could be interpreted as lack of ownership of the strategy by the executive team, and does not give the impression that risk management is very high on the agenda for the Trust Board. The Risk Management Strategy is, however, discussed at the Clinical Governance Committee, which is a sub group of the Board. The minutes and reports from this committee are presented to the Trust Board for information. Therefore, it could be argued that the Trust Board are at least informed about risk issues, even if they appear not to take a proactive stance on reviewing the success of the trust Risk Strategy objectives.

The Trust Risk Management Strategy sets out the responsibilities for risk management across the organisation at all levels. The overall impression resulting from the evidence collected is in this review is one of general understanding of risk management, although the managers demonstrated a lack of clarity regarding facilitating staff and ensuring effective two-way communication of risk assessments and incident report feedback. The lack of involvement in pro-active risk assessment process at all levels is concerning and reflects a possible lack of awareness of statutory requirements. There is a definite need for reinforcing roles and responsibilities as set out in the strategy. Staff need to be engaged with hearts and minds, empowering individuals to lead by example (Harris, 2000).

Both staff groups (A-E grades and managers) showed a lack of knowledge regarding the principles of risk management and indicated they wanted training. The application of risk management methods underpins a risk management strategy (Roberts, 2002), and there is evidence that the staff are applying some risk management methods by reporting incidents, but there is no consistency in the perception of feedback and sharing the lessons learnt.

Using the model described by Dale and Woods (2000), the findings from the survey have been presented to the Trust in the form of a management report, together with an action plan to provide feedback and information to the departmental managers. Part of the action plan will require a second cycle of the survey to be carried out within the same clinical area to demonstrate the effectiveness of the actions and that the trust Risk Management Strategy is functioning in a more effective way. Dale and Woods (2000) argue that any risk strategy must be implemented and facilitated by a coherent and manageable communication and evaluation process.

Future surveys should distinguish between trained and untrained staff to see if professional accountability impacts on risk awareness and behaviours. The medical team should be encouraged to participate in future studies to facilitate a multidisciplinary evaluation and whole team approach, as any changes implemented are more likely to succeed and be sustained if the whole team recognises the need for the changes.

This study represents a pilot for a future review of how the Trust Risk Management Strategy is functioning on a wider organisational level. At the local level, the review was found to raise local awareness of the trust Risk Management Strategy. A review across the whole organisation could, therefore, significantly impact the beneficial implementation of the Trust Risk Management Strategy.

References

Anderson P (2004). What is the impact of risk management in NHS trusts? *Healthcare Risk Report* Vol 10 Issue 3 p 12-13.

Dale C and Woods P (2000). A risk assessment and management strategy for community nursing. *British Journal of Community Nursing* Vol5 No.6 p 286-291.

Emslie S (2004). Why risk management is taking greater precedence at Board level. *Healthcare Risk Report* Vol 10 Issue 3 p 20-21.

Harris A (2000). Risk Management in Practice- How we are managing. *British Journal of Clinical Governance* Vol 5 No.3 pg 142-149.

Johnson G and Scholes K (2002). *Exploring Corporate Strategy: Text and Cases*. Prentice Hall, London.

Lugon M (2003). Avoiding the same mistakes, learning from incidents. *Clinical Governance Bulletin* Vol 14 No.3. The Royal Society of Medicine Press

Roberts G (2002). *Risk Management in Healthcare* (2nd edition). Witherbys Publishers, London.

Semple-Piggot C (2000). *Business Planning for Healthcare Management.* (2nd edition). Open University Press.

Wilson J (1997). Formulating a Risk Management Strategy. *British Journal of Healthcare management* Vol3 No.11 p 605-606.

3

The application of a risk management approach in the planning, development and commissioning of an NHS Walk-in Centre

LINDA CAMP

Introduction

This chapter outlines the risk management approach, based on the West Lancashire Primary Care Trust (WLPCT) Risk Management Strategy (2004), adopted in the planning, development and commissioning of the NHS Walk-in Centre in Skelmersdale, West Lancashire.

The concept of a Walk-in Centre is a relatively recent development in the National Health Service (NHS) and was originally announced by the Prime Minister at a health conference in Birmingham in April 1999. Walk-in Centres are complementary to existing primary care services and provide locally delivered integrated care over an extended working day and at weekends. Walk-in Centres are situated in locations convenient for patients, for example in shopping malls, high street retail outlets and airports. The Prime Minister opened the first centre in 2000 in Peterborough. A Walk-in Centre at Loughborough was also in the first wave and opened shortly afterwards.

In January 2004 there were forty two NHS Walk-in Centres already open when Health Minister John Hutton announced in the House of Commons that a further eleven were to be developed, one of which would be in West Lancashire.

The Walk-in Centre for West Lancashire is sited in a shopping complex in Skelmersdale town centre in what was an empty unit, which had previously been fast food outlet. It serves a population of 109,000 scattered across the rural area of West Lancashire. The Centre was opened by the Secretary of State for Health in November 2004 and provides open access to the public seven days a week from 07.00 to 22.00 Monday to Friday and 09.00 to 17.00 on weekends and Bank Holidays. The Centre is led by nurse practitioners and provides a minor injury unit and treatment rooms. Also provided at various times are dentistry, podiatry, phlebotomy, retinal screening, and a sexual health clinic. A community pharmacy and the general practitioner (GP) out of hour's service are also based at the centre. In 2005, radiology facilities and some pathology investigation capability was added to the Centre's range of services.

West Lancashire Primary Care Trust (PCT) has evolved from local reorganisation of health services and was established on 1 April 2001. The West Lancashire PCT is co-terminus with the boundary of West Lancashire District Council and the parliamentary constituency of Lancashire West, currently a Labour-held seat.

The PCT has a well- embedded risk management culture for existing services and staff. This chapter assesses the effectiveness of the risk management strategy with not only a new PCT service and staff but also a new NHS provision. The attached Executive Summary and Action Plan will be presented to the West Lancashire PCT Risk Management Committee.

Risk Assessment of the Walk-in Centre Project - October 2003

Senior NHS staff held a series of stakeholder meetings and a list of seven options were drawn up. An option risk appraisal was carried out using a standard 5x5 scoring matrix against 10 criteria. The option with the least risk was identified as the refurbishment of an empty fast food outlet. All of this information was provided within the West Lancashire PCT (WLPCT) business plan submitted to the Cumbria and Lancashire Strategic Health Authority for approval. Once approval was given and finance made available a further risk assessment of the project was carried out and a risk register established. The identified high risks were:

Financial
- Insufficient management capacity to manage large capital programme
- Capital programme exceeds capital resource limit
- Building not handed over on schedule
- Risk to revenue with high patient demand

Clinical
- Inability to recruit and retain highly qualified staff
- Patient expectation and demand outstrip resources
- Patient demand does not match projection (under use)
- Inappropriate use of the Walk-in Centre in respect of Mental Health clients

Reputation
- The PCT fails to meet its objectives
- The PCT fails to deliver on the NHS plan

Political
- Local Health campaign group dissatisfied with outcome
- Failure to open on schedule as Secretary of State has agreed to open centre
- Breach of Statutory Duty

This further risk assessment was conducted against a robust framework and through informal interviews with key staff together with conducting a review of pertinent documentation from within the PCT and also, for comparison, from the commercial sector.

The staff interviewed included the Chief Executive, Director of Service Delivery, Director of Partnership and Public Participation, Director of Corporate Services, Head of Finance, Head of Information Management & Technology, Head of Estates, Walk-in Centre Manager and deputy and Walk-in centre staff.

The framework that was adopted was adapted from the European Foundation for Quality Management (EFQM) model (www.efqm.org). The EFQM model was created in 1989 as collaboration between 14 leading European businesses that agreed on the essential components for achieving excellence in an organisation. It is now used in 19 European countries by many organisations in the achievement of high quality outcomes by involving all staff in improving processes. The model enables an organisation to identify not only good practice but also areas for improvement. The theme of the model is the alignment of strategy with the results sought and the approaches deployed. There is also a focus on partnership working, again an emerging theme within the NHS. There is a requirement to provide evidence to support the assessment process. As part of a Department of Health pilot project, eight senior managers at West Lancashire PCT were trained as EFQM assessors. The risk management strategy was assessed against the seven key areas of the EFQM model in figure 3.1

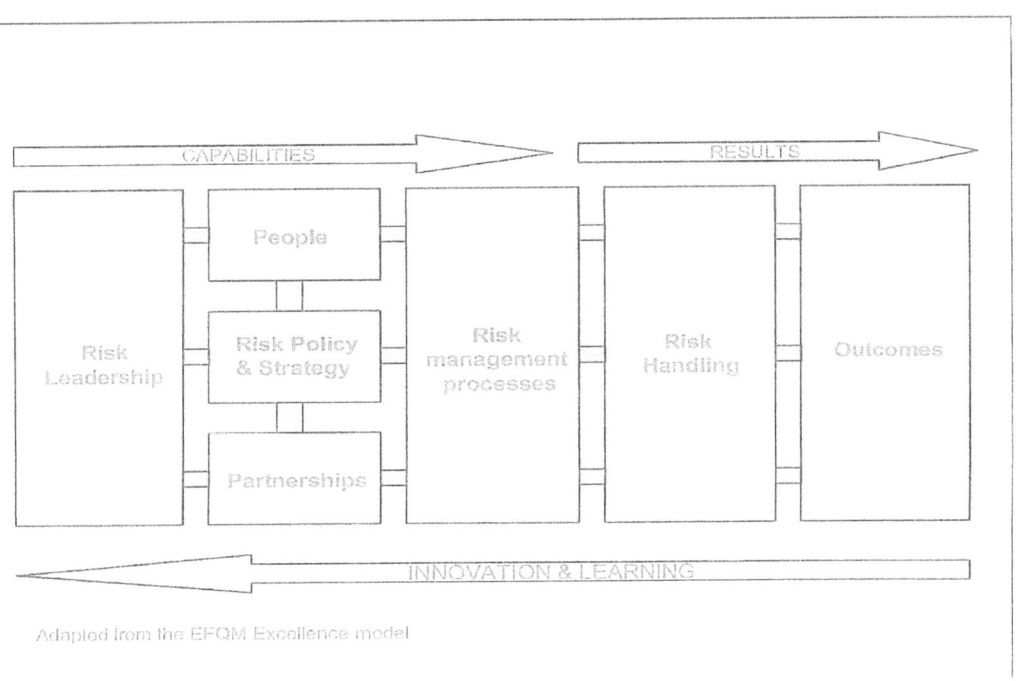

Figure 3.1 – EFQM model adapted for healthcare

1. Risk Leadership

Roberts and Jolly (1996) consider that "A risk management strategy and programme is a sign of commitment by a healthcare organisation to quality and excellence". The WLPCT Risk Management strategy states that "The board believes that risk management is a normal part of the governance process and continuous quality improvement." Board level commitment is essential for risk management to succeed in any organisation. This is the underlying principle of the Turnbull report (Institute of Chartered Accountants in England and Wales, 1999). The Turnbull report advocates a holistic approach to risk and control across the entire organisation. This is particularly relevant to the NHS where risk had traditionally been compartmentalised into Health and Safety, Clinical and Financial risk, and rarely had reputation risk been considered. West Lancashire PCT has from its establishment adopted a holistic approach to risk management. There is a single Risk Management committee chaired by a non-executive director. The Chief Executive, Trust Chairman and Directors of Finance and Clinical Services are committee members.

Evidence of Good Practice

An executive Steering Committee was established for the Walk-in Centre. This committee was chaired by the Chairman of the PCT, met monthly and received update reports on estates, financial expenditure and cost pressures and clinical development. The committee was responsible for resolving issues at a strategic level. It reported directly to the trust board. There was also a Project Board which was a decision making board to ensure the project milestones were completed on schedule. There were three operational subgroups of this board considering Clinical Service development, Information, and Technology. There was also a Stakeholders group. The sub groups reported to the Project Board. All groups had documented Terms of Reference. Project Board minutes and trust board minutes were used as evidence that these arrangements were sound. A risk log with control measures had been established and the Project Board was fully informed by regular update reports of any uncontrolled risks.

Leadership and understanding of the risks, whilst not stifling innovation, was evident at all levels of the project with highly skilled professionals leading in many cases, small teams to ensure the project was successful.

Areas identified for Improvement

Not all staff interviewed were able to identify the project manager. The project manager needed to be clearly identified to everyone involved.

2. Policy and Strategy

The West Lancashire PCT risk management strategy is reviewed annually and is a board-approved document. This strategy supports the WLPCT Service Strategy (2002-2007). The service strategy was developed in conjunction with staff and stakeholders over a 12 month period from the trust's establishment in 2001. The risk management strategy identifies significant risk as anything that prevents the organisation from achieving its objectives, in this case the opening of the Walk-in Centre. The Australia/New Zealand risk management Standard (Standards Australia, 2004) was adopted across the NHS and this helped to shape the PCT risk management strategy.

Evidence of Good Practice

The PCT scored 100% for its risk management strategy against the NHS Litigation Authority PCT risk management criteria (www.nhsla.com).

The PCT has more than 100 policies to support the risk management strategy. All PCT strategies and policies are available to staff and the public via the PCT web site (www.westlancspct.nhs.uk).

Risk was further mitigated through the production of standard operating procedures (SOPs) specifically for the Walk-in Centre. Around 100 SOPs were been developed in advance of the Walk-in Centre becoming operational.

Areas identified for Improvement

Further standard operating procedures needed to be developed once the Walk-in Centre was operational.

The trust needed to be assessed against more stringent aspects of the NHSLA risk management standard in 2005.

3. People

Residents of Skelmersdale had actively campaigned for a hospital in the district and a campaign group had been in existence for a number of years. The campaign was never going to be achieved as there were already three District General Hospitals within a 10 mile radius of Skelmersdale. The Trust had a Director of Partnership and Public Participation, who fully engaged with the campaign group and residents and secured their support for the Walk-in Centre. The residents of Skelmersdale were invited to presentations and their opinions sought.

A Walk-in Centre Manager was appointed in May 2004 to manage the Centre on a daily basis; she reported to the Director of Clinical Services both through formal reports and almost daily meetings to ensure that the centre opened on schedule. A further 25 staff were appointed to work in the Walk-in Centre.

According to Reason (1997) human failures rather than technical failures often pose the greatest threat to complex systems, and unfamiliarity with a task is often the highest risk factor. To help minimise this risk, a full team of staff, who had all been appointed externally to the PCT specifically for the Walk-in Centre, commenced employment in phases and were all in post by October 2004 to enable team building, training and task familiarisation to commence. Across the organisation almost 40 senior staff were involved in the development of various clinical specialities and support services.

Evidence of Good Practice

A Stakeholder sub group was fully involved in the development of the Walk-in Centre. There were regular meetings that were minuted. There was Stakeholder representation on the Project Board.

Highly qualified nurse practitioners lead the services in the Walk-in Centre. There is evidence of clear job descriptions for the practitioners and all staff have personal development plans in place.

Corporate induction and other training, including clinical procedures e.g. resuscitation and proper handling of controlled medicines, and security (both personal and site) were ongoing for some months before the Centre's opening. Staff who were interviewed were enthusiastic about their new roles and felt well supported in this new venture.

Areas identified for Improvement

Further training was required to familiarise staff with emergency procedures in relation to the shopping concourse.

4. <u>Partnership and Resources</u>

The NHS (Primary Care) Act 1997 charges Primary Care trusts with three key functions: To achieve financial balance; to improve the health of the local population; and to reduce health inequalities.

The Act also gives much more freedom to PCTs, than was previously available, to work in partnership with local authorities, the private sector and other public organisations. The Walk-in Centre required significant partnership working with public, private and voluntary organisations. As a major direct employer in the area the PCT endeavoured to use local companies wherever possible to carry out the development.

Evidence of Good Practice

Partnership working was evident with local NHS trusts and with Lancashire County Council, West Lancashire District Council, the GP out of hours service, the property company that owned the shopping concourse where the Walk-in Centre is sited, and many voluntary agencies. There was, and remains continuing involvement of all the main stakeholders. There has been close dialogue with General Practitioners to gain their support for enhanced services.

Areas identified for Improvement

A support group for the Walk-in Centre is to be developed. The group must adhere to sound governance principles.
Any equipment either donated or purchased must be fit for purpose.

5. <u>Processes</u>

Within the risk management strategy there is a single process for identifying, analysing and evaluating risk across the organisation. Each directorate has a risk register which feeds the corporate risk register. 'High' risks are reported quarterly to the board via a risk register. A risk register was produced for the Walk-in Centre with control measures clearly identified. All staff had undergone mandatory training prior to the centre opening.

Evidence of Good Practice

The Business plan contents page shown in figure 3.1 is provided as evidence of good practice in that risk was considered from the outset of the project. Minutes of the Walk-in Centre Steering Group contained a copy of the risk log developed by the IM&T manager. An operational risk register was developed by the Walk-in Centre manager and was included in the steering group minutes.

1.0	Introduction
2.0	Executive Summary
3.0	The Strategic Content
4.0	The Case for Change
5.0	Objectives, Targets and Constraints
6.0	Qualitative Option Appraisal
7.0	Financial Evaluation of Options
8.0	Option Appraisal of Risk
9.0	The Preferred Option
10.0	The Preferred Option Facility & Service Content
11.0	The Case for X Ray provision
12.0	Timetable for delivery
13.0	Benefit Realisation & Post Project Evaluation
14.0	Project Organisation

Figure 3.1 – Skelmersdale Walk-in Centre: Business Plan Contents page

Areas identified for Improvement

Concern was expressed that some of the risks were outside the control of the directly employed PCT staff and mechanisms needed to be found for addressing these risks. Whilst the risk registers had been prepared for various aspects of the Walk-in Centre project there was no overall co-ordinated risk log. The risk registers indicated high risks that had not been reported to the trust risk management committee. There were occasions where weakness in the communication between the various sub groups was identified as a key risk.

6. <u>Risk Handling</u>

Risk is inherent in life; the capacity to handle risk is often a confidence issue rather than a matter of competency. All staff attend a mandatory risk management training programme and additional training is provided to staff appropriate to the job and position within the organisation.

Evidence of Good Practice

Forty senior managers, including the Chief Executive, have passed the Institution of Occupational Safety and Health's "Managing Safely for Health Care professionals" (www.iosh.co.uk). A further 40 staff have passed a certificated risk assessors course.

Areas identified for Improvement

New in post senior staff need were identified as needing access to relevant risk management training. A risk self -assessment workshop was planned for all new staff in the Walk-in Centre. It was considered that this would provide an opportunity for new staff to voice concerns and to check out processes.

7. <u>Outcomes</u>

The outcome of the project was a million pound state of the art Walk-in health centre. Its impacts may be considered in relation to meeting the five key objectives of the Walk-in Centre, i.e.:

1. improving access to healthcare services for the population of Skelmersdale;
2. contributing to the reduction in health inequalities in Skelmersdale;
3. contributing to reducing waiting times for patients to see existing primary care health professionals;
4. contributing to reducing unnecessary use of A&E and Minor Injury Unit hospital-based services by patients with very minor complaints; and
5. providing facilities that enable patients to be treated or seen in a close to home primary care setting rather than at an acute hospital.

Conclusions

The Walk-in Centre has provided the people of Skelmersdale with a much needed health facility, providing patients with a local easily accessible extended hours NHS service. This was particularly important for the local population as the local adult Accident and Emergency department was transferred in 2005 to a district general hospital 14 miles away at Southport. The shopping concourse has 60,000 people a week using the centre; there is now a readily accessible health facility. An initial evaluation of Walk-in Centres by the University of Bristol for the Department of Health showed that 50% of attendees access the service as a substitute for a GP consultation and that working age males are more likely to access a Walk-in Centre rather than arrange a GP consultation, thereby improving access to primary care for a group of 'difficult to reach' section of society.

For the staff the opportunity for professional advancement as nurse practitioners and career opportunities has enabled the PCT to recruit high calibre staff both locally and nationally.

As a result of the assessment identified in this chapter, there now exists a risk aware culture within the PCT. Ares identified for improvement have been acted upon. Overall, the trust risk management strategy has been effective in supporting the Walk-in Centre project, ensuring objectives were met and good outcomes achieved.

References

Standards Australia (2004). *Risk Management*. AZ/NZS 4360:2004. www.riskmanagement.com.au

European Foundation for Quality Management. European Foundation Excellence Model (Public and Voluntary sectors). www.efqm.org

Institute of Chartered Accountants in England and Wales (1999). *Internal Control - Guidance for Directors on the Combined Code (the Turnbull Report)*. www.icaew.co.uk/internalcontrol.

NHS Litigation Authority (2004). *Risk Management Standard for Primary Care Trusts*. May 2004. www.nhsla.com

Reason J (1997). *Managing the risks of organizational accidents*. Ashgate Publishing.

Roberts G and Holly J (1996). *Risk Management in Healthcare* (1st edition). Witherbys Publishers, London.

West Lancashire PCT (2004). *Risk Management Strategy*. www.westlancspct.nhs.uk

This page is intentionally blank.

4

Implementation of a risk management strategy within the Chronic Disease Management Team of a Primary Care Trust

EMILY HACKETT

Introduction

The Government White paper *the New NHS: Modern and Dependable* contained proposals for the introduction of a system of clinical governance in all NHS Trusts (Department of Health, 1997 para. 3.6). This was reinforced by HSC 1999/123 *Governance in the new NHS: Risk management and organisational controls* (Department of Health, 1999), which linked clinical governance and the assurance of quality clinical services to underpinning financial and organisational control systems through the "common thread of risk management." HSC 1999/123 asserts that the achievement of sound systems of risk management and financial/organisational control forms a solid foundation on which excellence in clinical care can flourish (NHS Executive, 1999, p.3). Crucial to sound risk management and internal control is the direction and guidance that is provided by an organisation's risk management strategy.

This chapter evaluates the effectiveness of the risk management strategy within a service delivery area of an NHS Primary Care Trust (PCT), examining how the strategy is communicated and applied, and identifying any weaknesses in the systems described within the strategy that can be considered and amended to strengthen the risk management process as a whole.

The Primary Care Trust

Newark and Sherwood PCT is located in North Nottinghamshire, serving a population of around 125,000 in predominantly rural settings.

The PCT manages the following direct patient care services:

- District nursing
- Health visiting

- Intermediate care – providing 24 hr nursing care to prevent hospital admission and promote early discharge from hospital
- Chronic Disease Management – providing support for people with long term conditions
- Community Child Health Nursing
- Balderton Medical Practice
- Drug and Alcohol action team
- Hetty's/WAM – a support service for families of drug users and (What about me) support for young persons worried about someone else's drug use
- Smoking Cessation service (Newark and Sherwood PCT 2003, p7)

The implementation of the risk management strategy is, for the purposes of this chapter, examined in the context of the Chronic Disease Management team. The Chronic Disease management team was formed in 2004 following recommendations from the Department of Health (Department of Health, 2004 p.35). In observing the risks the team faces and the steps it takes to control them, the risk management strategy and policy will be reviewed in its widest perspective, taking into account the context of the clinical governance and risk management agendas within the NHS and the national requirements for the content and application of the risk management strategy in terms of the NHS Litigation Authority standards and the requirements both of controls assurance standards and the new Standards for Better Health.

Definitions

In its publication 'Management of risk – guidance for Practitioners' the Office of Government Commerce (2002) looks at the framework that organisations should put in place in order to be able to take informed decisions about risk. It looks at the structure for risk management at strategic, programme and project levels; at each level offering an explanation of the policy requirements for the management of risk, its composition and the responsible parties. The term "policy" is used by the Office of Government Commerce to describe the document or statement of how risks will be managed throughout the organisation or through the lifetime of the programme or project.

The NHS Litigation Authority (NSHLA) sets requirements on the documentation that NHS organisations use to define its risk management direction. A key criterion in the NHSLA's risk management standards is "The organisation's senior management has defined and documented its strategy for managing risks, including objectives for, and its commitment to, risk management. The risk management strategy is relevant to the organisation's strategic context and its goals, objectives and the nature of its business. Management ensures that the strategy is understood, implemented and maintained at all levels of the organisation." (NHS Litigation Authority, 2004)

Within Newark and Sherwood PCT, the documentation is clearly defined within the trust's 'Process for Adoption and Ratification of Policies and Procedures' (Newark and Sherwood PCT, 2003). This guidance sets out the organisation's definitions and guidance for writing strategy, policy and procedure. For the PCT, the Board level document is the Strategy; the Policy outlines the organisational direction based upon that strategy; and the procedures give detailed guidance on how to achieve the objectives of the policy.

The key lead document for the management of risk within the PCT is the 'Risk Management Strategy and Policy 2004'. It includes the strategic statement of intent and organisation as well as responsibilities, governance arrangements and a step-by-step guide for all staff to the risk management process. It was developed across the PCT, with contributions from all directorates, and consultation with staff at all levels through the committee structure, to confirm it could be understood and applied across the organisation, and to increase ownership of the document and the process it describes for all staff.

Risk management process

In common with many NHS organisations, the risk management process adopted by the PCT is based on the Australia/New Zealand Risk Management Standard (Standards Australia 2004), which was recommended for use within the NHS by the Department of Health's Controls Assurance Team (Department of Health, 1999 p.4). The Risk management strategy and policy was, in addition, written to meet the requirements of the NHS Litigation Authority's risk management standard for PCTs (NHS Litigation Authority, 2004).

Risk Management Strategy

The role and function of the risk management strategy within organisational management has its roots in the management of Health and Safety. The Health and Safety at Work etc. Act 1974 requires that a policy on health and safety must exist within an organisation and be recorded if more than five people are employed. In more composite organisations the requirement of the policy and the risks that it intends to manage has a wider scope. Jeyes (2002) breaks the risk elements that face an organisation into four distinct groups under the heading 'management issues': policy, strategy, planning and organisation.

Waring and Glendon (1998) draw on examples of risk management and control failure from all types of organisations around the world. They espouse the need for a holistic approach to risk management. One example they cite is the collapse of Barings bank. The breakdown of the control failures in place can be reviewed and compared with the risks faced by equally complex NHS organisations. In this way, sharing the lessons learned from previous organisational failure prevents NHS organisations from being insular in their approaches to management controls and the risk management processes in place. This approach is discussed further by Walshe (2003), who believes that if healthcare systems had better mechanisms for identifying, investigating and learning from major organisational failures, they would almost certainly be better at preventing such failures in future.

The 'Turnbull Guidance' (Institute of Chartered Accountants in England and Wales, 1999) includes explanations of the policies regarding internal control and the factors that should be considered when assessing the risk management system. Turnbull confirms that it is the role of management to implement board policies on risk and control and explains that an internal control system encompasses the policies, processes, tasks, behaviours and other aspects of a company. Using this approach makes the risk management strategy a holistic document, looking at not only the processes for risk management, but also organisational culture, responsibilities and the underpinning procedures and task guidance. This is further explained by Hopkin (2002, p.122), who believes that the risk management policy should facilitate successful implementation of enhanced risk management in the organisation.

Many of these areas are now requirements for inclusion in a risk management strategy within the NHS, as originally required by the controls assurance project and the NHS Litigation Authority. All NHS organisations must include in their strategy definitions of all levels of accountability and descriptions of key responsibilities for staff within the risk management strategy in order to achieve compliance with national standards. The NHS Litigation Authority Risk Management Standard (2004) states that it must include an illustration of the risk management organisational structures, an outline of the risk management process and the assessment tools in use in the PCT. It also contains requirements for the communication of the strategy. These are all areas included within the current risk management strategy and policy that will be considered as part of this review.

HM Treasury (2001) discusses an embedded risk management process stating that at every level of objectives there should be a parallel delegation of responsibility for the associated risk issues. The PCT risk management strategy and policy follows this concept, detailing the responsibilities for risk management of the PCT Board, directors, managers and all staff. It also details the responsibilities of specialist advisors for fire and health and safety management that are purchased through service level agreements with neighbouring NHS organisations. The NHS Litigation Authority (2004) further reinforces this sentiment when it says that the management of risk should be integrated into the management philosophy of an organisation.

The examination of the effectiveness of the risk management strategy and policy will focus on certain key areas, one of which is incident reporting and management. Gale (2002) believes that one of the main challenges for risk managers within the NHS is to work with managers and staff to promote a more open and supportive culture, to encourage staff to report incidents and near misses and to enable the organisation to learn from these. The National Patient Safety Agency (NPSA) (2004) advocates the development of local risk management systems that are designed to help NHS organisations manage incidents effectively within its integrated risk management activity. The PCT risk management strategy and policy includes a statement of organisational culture and is linked and to the incident reporting and management procedures.

The building of a safety culture is step one of the NPSA's seven steps in its 'seven steps' guide to patient safety. The Agency encourages trusts to ensure that policies and procedures are in place and that all staff feel able to discuss concerns and report incidents. The review set out below aims to show how effective the Newark and Sherwood PCT has been in promoting this throughout the organisation, in the context of the Chronic Disease Management Team.

Methodology

In order to assess the effectiveness of the risk management strategy and policy within the Chronic Disease Management Team, a questionnaire was developed to gather data on six key areas.

1. **Risk management strategy and policy** – focusing on general awareness of the policy and the responsibilities described within it.

2. **Policies and procedures** - focusing on the awareness, understanding and

effectiveness of communication of the policies and procedures underpinning the risk management strategy and policy.

3. **Training** – focusing on communication, accessibility and appropriateness of the risk management and general training available to staff.

4. **Incident reporting** – looking at the organisational culture around reporting and feedback on reported incidents.

5. **Risk register** – focusing on staff awareness of reporting risks and the function of the risk register.

6. **Risk management** – allowing staff to offer their views on the effectiveness of risk management, how they would influence the process and how they believed risk management within the organisation helps to deliver the services.

Guidance on questionnaire design in the context of risk management was taken from Dickson (2003). In order to facilitate ease of analysis of the collected questionnaire, a grading system was devised for the closed questions which could be converted into numerical responses. This system looked at a series of positive statements and asked the respondent to mark their level of agreement to the statement. The level of agreement had a corresponding numerical value that could then be analysed. Open questions were also included, allowing the respondent to answer in their own words. This cross method approach of using qualitative and quantitative questioning allowed the respondents a wider choice of answers and gave wider scope for analysis of the data collected.

Informal discussions were also held with the participants, allowing the feelings of the staff towards the risk management process to be voiced. The effectiveness of the strategy was measured again using the six key areas and staff awareness and understanding of those areas and their responsibilities under them.

Analysis and Discussion

Eleven questionnaires were distributed with seven being returned. Nine members of the team were present for the discussion and participated well, asking questions and offering opinions.

1. <u>Risk management strategy and policy</u>

The results showed a good awareness of and accessibility to, the strategy and policy. In general, staff awareness of their own and others' responsibilities under the strategy and policy are known, with six of the seven agreeing that they felt familiar with the policy and responsibilities. All respondents agreed that they understood the risk management process as described within the policy.

This showed that the communication of the policy to staff within this team had been successful. The policy includes a step-by-step guide to the risk management process that is being prepared for a booklet for all staff following wide staff consultation. The results of this section show that this, and the initial consultation during the development of the strategy, have been successful. Work in this area must continue as new staff join, and any changes or reviews of the strategy need to be captured and communicated. West (2001) asserts that participation appears to be most effective when it is a permanent and inclusive feature of the employment relation rather than sporadic or exclusive. In considering this, the PCT must continue to include employee participation in the development of all of its risk management policies and procedures in order to create greater awareness and ownership.

2. <u>Policies and procedures</u>

The results for this area showed a more varied response than the previous section. Here staff were asked about their awareness and understanding of underpinning policies and procedures including the health and safety policy, incident reporting procedure and manual handling policy – see figure 4.1.

The results here show that the general awareness of the policies and procedures listed was high. The results do, however, highlight an anomaly with respondent 3, who appears to have an understanding of the policies and procedure, but no awareness of them! This is most likely an error on the part of the respondent in responding to questions on awareness.

In order to increase understanding of these key policies and procedures, the PCT must address policy training and review participation as a consistent part of the policy development process.

	Awareness of policies and procedures (P/P)									Understanding of policies and procedures (P/P)								
	Respondent number									Respondent number								
P/P	1	2	3	4	5	6	7			1	2	3	4	5	6	7		
H&S	1	1		1	1	1	1	6		1	1	1	1	1			5	
HV	1	1		1	1	1	1	6		1	1	1		1			4	
IR	1	1		1	1	1		5		1	1	1		1		1	5	
II	1	1		1	1	1	1	6		1	1	1		1			4	
MH	1	1		1	1	1		5		1	1	1	1	1		1	6	
SS	1	1		1	1	1		5		1	1	1		1		1	5	
V&A	1	1		1	1	1	1	6		1	1	1	1	1			5	
WM	1	1		1	1	1	1	6			1	1	1	1			4	
Clin	1	1		1	1	1	1	6		1	1	1	1	1			5	
Total	9	9	0	9	9	9	6	51	81%	8	9	9	5	9	0	3	43	68%

Key to policies and procedures (P/P)

H&S	Health & safety policy
HV	Home visit guidelines
IR	Incident reporting procedure
II	Incident investigation procedure
MH	Manual handling policy
SS	Staff support policy
V&A	Management of violence and aggression
WM	Waste management policy
Clin	Clinical policies and procedures

Figure 4.1 – Results for policies and procedures

The team were also questioned on their preferences for communication of policies and procedures. One respondent stated that the issue was one of "bringing it to the troops, rather that just issuing briefings". This is reflected in the preferences shown in figure 4.2, the largest portion showing that team briefings are the favoured method of communication.

These results mirror the results of an internal questionnaire on communication completed by all staff in June 2004 for the *Improving Working Lives* initiative (www.dh.gov.uk). It found that team briefing was preferred by 65% of respondents; and 75% felt that the information could be relied on as being more accurate.

A concern for the PCT to take on board is the lack of awareness of the *Health Safety and Risk Control Workbook* as a method of communication. This is a reference tool that is used throughout the PCT, giving information on all health and safety issues and reference to policies, procedures and guidance. The lack of mention of this tool may be due to the relative newness of the team within the PCT.

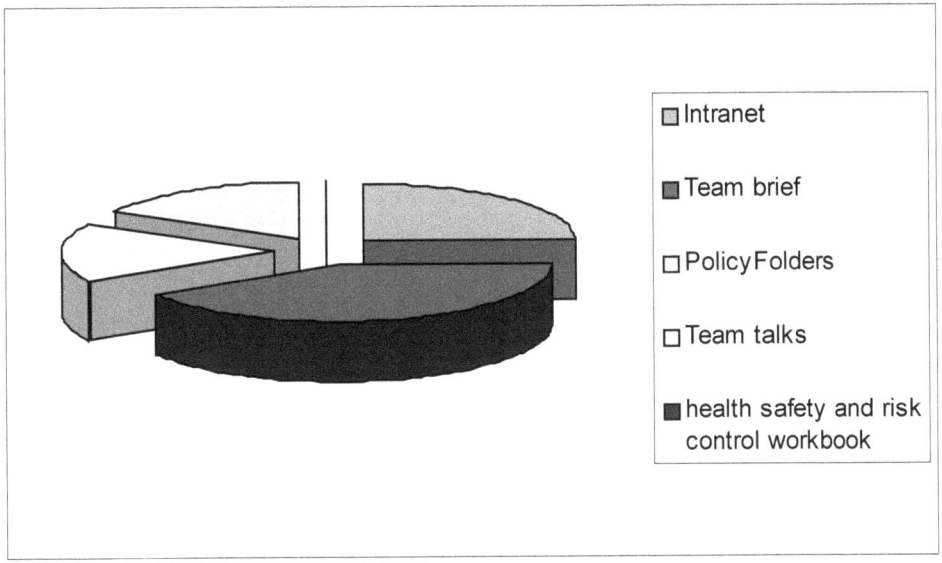

Figure 4.2 - Preferences for communication of PCT Policies and Procedures

3. Training

The questionnaire also looked at training, focussing on both awareness and availability. Six respondents agreed that they were aware of the training available including the mandatory training responsibilities. The respondents agreed that training issues were well communicated throughout the PCT, however two of the respondents did not feel that there was enough training in the areas they felt were important for their job. One reason for this may be that, as the team develops, its training needs will be identified and the provision of training will need to be sought to match those needs, but that currently these training requirements have not been sufficiently identified. One risk area that was highlighted by the team, and that is already identified within the PCT, is the time taken to access training. This is an issue that is being considered jointly by the PCT managers and the providers of the training for the PCT

4. Incident reporting

The team showed here that they have good awareness of how to report incidents and also of the follow-on incident management process. This is an area where the PCT has worked in line with national initiatives and in conjunction with the NPSA regarding the reporting culture. Staff here agreed, and two strongly agreed, that the culture within the PCT encourages them to report incidents and near misses and that there is support from managers following incidents. This is also shown in the incident reporting figures, as a slight increase has been seen quarter on quarter since 2002.

Within this time, the range of incidents reported has also widened, with staff feeling able and supported to report incidents involving drug error and other clinical issues. The PCT has looked into the increase in reporting figures and has determined that it is as a result of the increased awareness and support mechanisms that are in place. Emslie (cited by Anderson, 2004 p. 13) states that he would "steer clear of making a meal of linking risk management to a reduction in numbers of incidents, but instead focus on learning from their analysis and on other outcome indicators." Within the PCT, this has been translated into the incident investigation processes and training in root cause analysis to learn from incidents and continue to encourage their reporting; this needs to continue. The suggestion was made that both the process for reporting and the reporting form used should be simplified; this is an areas that the PCT must consider in the future.

5. Risk register

The team showed here that they are not all aware of the mechanisms in place to report risks to the organisation and that they do not feel empowered to influence the reporting of risks. More than half stated that they do not know whom they would talk to about reporting risks to the organisation. For the PCT this is a large area of concern. The step by step guide to the risk management process within the PCT is being developed and will be available for all staff in booklet form. Alongside this, the PCT will need to raise awareness of risk reporting, the mechanisms in place and the risk assessment tools that are available for all staff.

This review has shown that this area is in need of development and the PCT needs to put a plan together too look at this. This area may be weak for the PCT, as it relies on the health, safety and risk control work book for much of its health and safety risk reporting. This group has already shown to not judge this tool as a useful one.

6. Risk management

This section of the questionnaire allowed the team to comment on what they felt risk management had contributed to the delivery of services within the PCT. In response to the question "what is risk management?", the comments were mainly ones of ensuring safety and support for staff, assessing risks and reducing the incidence of harm to patients.

To the question "what changes in risk management activity had they witnessed?", the range of responses was wider, ranging from 'no changes seen', to the 'introduction of post incident support' and a 'cultivation of a 'no blame' culture across the PCT'.

The change that respondents desire to see is to demonstrably bring risk management skills to all staff, rather than simply issue briefings. Team risk management discussions were felt to be a worthwhile activity and an effective way of communicating issues which allowed staff to ask questions and have instant feedback.

The final question looked at ways the team would like to influence processes within the PCT. For risk assessment, they felt that training was the most important issue. For incident reporting, the documentation was highlighted as being an issue that needs addressing, with the team wanting a simpler form and reporting process. For risk communication, the issue raised was one of sharing risks and lessons learned across both the PCT and the local health community.

Graded questions

The questions that had a grading system with a numerical score can be looked at separately. Figure 4.3 identifies the questions. Figure 4.4 shows the responses. Whilst most responses are very encouraging, with respondents principally 'agreeing' or 'strongly agreeing', it is interesting to note that the question scoring most highly is the 'culture' question (19).

No.	Question description
4	I feel familiar with the Risk management strategy and policy and its contents
5	I understand my responsibilities under the Risk management strategy and policy
6	I understand my managers responsibilities under the Risk management strategy and policy
7	I understand the risk management process as described in the Risk management strategy and policy
11	I am aware of the training available to me
12	I am aware of my responsibilities under the training policy with regards to mandatory training
13	I feel that Training issues are well communicated within the PCT
18	I obtain adequate feed back from the incident/near miss reports that I have submitted.
19	The PCT culture encourages me to report incidents/near misses of all kinds
20	My manager supports me following an incident/near miss

Figure 4.3 – Graded questions

Question Number	Strongly Disagree (1)	Disagree (2)	Don't know (3)	Agree (4)	Strongly Agree (5)	Total
4		1		6		26
5			1	6		27
6		1		6		26
7				7		28
11				6	1	29
12				6	1	29
13				7		28
18		1	1	5		25
19				5	2	30
20			1	5	1	28

Figure 4.4 – Graded questions: responses

Concluding comments

This chapter has presented the author's approach to conducting a review of the implementation of the trust risk management strategy in terms of one of the service delivery areas – the Chronic Disease Management Team. Whilst the results of the review are encouraging, they may not be generalisable across the organisation. It would be useful to treat the current review as a 'pilot' and conduct the study across all delivery areas of the trust to give a complete picture of the effectiveness of the risk management strategy and policy.

The risk management strategy and policy is a key document within the governance arrangements of any NHS organisation. This review has shown that awareness and understanding, and, by inference, the effectiveness of the risk management strategy within the Chronic Disease Management Team is reasonable, but could be significantly improved. If these results are generalisable across the trust, then they would be encouraging. But more needs to be done. Appendix 4.1 contains a board report and action plan submitted to the Newark and Sherwood PCT board in response to the review in this chapter.

Some areas are functioning well for Newark and Sherwood PCT, the incident reporting processes are well understood and employees feel supported when reporting incidents. Risk assessment and reporting to the risk register is not functioning well. The process as described within the risk management strategy and policy has not been well communicated and staff have not been trained to enable them to carry out the requirements of the process.

Communication of policies and procedures, and staff awareness and understanding of them, is also varied. The *Improving Working Lives* group previously surveyed staff on their preferred and relied-upon method of internal communication, which was found to be team briefs, The PCT must expand the team brief to include communication from all directorates.

The review did show that, although the risk management strategy and policy meet basis national requirements in terms of governance and NHS Litigation Authority needs, they may not be meeting the higher standards set down by the PCT itself. Acceptance and implementation of the board report and action plan at Appendix 4.1 would significantly improve the effectiveness of the trust risk management strategy and policy.

Appendix 4.1 – Board report and action plan

Title of paper	**Review of risk management strategy and policy**
Meeting Date	
Person Presenting Paper	
Purpose of paper	**For Consideration**

Brief

A review of the effectiveness of the risk management strategy and policy was undertaken within the Chronic Disease management Team. This is a new service delivery area for the PCT; and it was decided that this team be used as a benchmark for all areas of the organisation

The purpose of the review was to highlight areas for development within the risk management processes, where the effectiveness of the strategy and policy was low and to confirm that areas of work that are underway match staff needs

Methodology

The methodology used was a survey of staff looking at six key areas of the strategy.

1. Risk management strategy and policy
2. Policies and procedures
3. Training
4. Incident reporting
5. Risk register
6. Risk management

Results

The results of the review showed that the PCT needs to continue to raise awareness of the strategy and policy for all staff. The preferred method for doing this is an increase in staff involvement in the review of all policies and procedures - as a permanent and inclusive process. This involvement in policy review needs to be expanded to take into account nursing forums, allowing the policy review to access larger staff groups.

It is also recommended that the Team Brief should be used more widely to communicate policies and procedures. This is in line with the 'Improving Working Lives' (IWL) review earlier this year which showed the Team Brief to be the preferred method of internal communication for staff. A review of the Team Brief, its format and effectiveness should be considered.

It also became apparent that the Health, Safety and Risk Control Workbook required review of format, content and use with in all PCT settings. This also applies to the assessment tools in use within the workbook and communication of the workbook.

Staff demonstrated that they felt that the PCT offered them a supportive culture in which to report incidents and they were aware of how to do this; promotion of incident reporting should continue; and further links should be created to the investigation process. Staff also requested a simpler reporting process and form. Under training, staff expressed the need to examine access to training, making it more timely and relevant to their needs.

Under general risk management issues, the need for awareness raising for all staff was expressed. The step-by-step guide being produced currently will go some way to explain the risk management process for all staff.

Recommendations

It is recommended that the attached action plan be adopted and regularly reviewed by the Operational Risk Management sub group.

It is also recommended that the review be reproduced and extended to a larger staff base, enabling more data to be gathered on opinion of the key areas and to assess the wider effectiveness of the risk management strategy and policy and the processes it describes.

Action Plan

It is recommended that the action plan is considered quarterly at the Operational Risk Management sub group and that the review process is repeated within 12 months using a larger group of employees.

The action plan should then be updated with the progress and any new areas that are highlighted.

Key area	Action	Lead	Timescale	Progress	Sources of assurance
Area highlighted by the questionnaire	*Action to be taken to address the issues highlighted*	*Lead for progressing action*	*Timescale for review of progress*	*Progress to date of actions*	*Measurable sources of assurance that actions are progressing successfully*
Risk management strategy and policy	Continue to raise awareness of the strategy and policy for all staff	Risk co-ordinator	On-going	<ul><li>Awareness good.</li><li>Step by step guide being produced</li><li>Review of all policies and procedures for FOI and as part of intranet project</li></ul>	<ul><li>Audit use of risk management process and assessment materials.</li><li>Re-evaluate questionnaire in 12 months over wider scope</li></ul>

Key area	Action	Lead	Timescale	Progress	Sources of assurance
Policies	• Increase staff involvement in the review of policies and procedures - as a standard and ongoing process • Involvement in policy review to be expanded to take into account nursing forums – and larger staff groups • Look into possibility of using consultation tools currently used for public and patient involvement,	Risk co-ordinator. PCT senior managers		Staff currently consulted using the Health and Safety and Committee and Operational Risk Management Sub Group	Greater involvement from staff in policy development Greater understanding and awareness of all policies and procedures from all staff groups
	Communication of policies and procedures using Team Brief	All Managers	Feb 04	Policy and Procedure review underway with a view to creating risk management policy folders and updating the intranet.	All staff have access to all up to date policies and procedures and team brief is used to communicate policies and procedures

Key area	Action	Lead	Timescale	Progress	Sources of assurance
	Health and safety and risk control work book review. Looking at format, content and use within all PCT settings.	Risk Co-ordinator	July 05		Wider awareness and usage of the workbook. Workbook contents in line with staff requirements
Training	Look into access to training	Deputy Director of Governance	April 05	Meeting set up with Training and Development managers to review access to mandatory induction training and its content	Staff able to access training within timescales requested
Incident reporting and management	Continued promotion of incident reporting. Integrate investigation process and facilitate simpler reporting processes	Risk Coordinator All managers	May 05	• Root cause analysis training underway • Datix risk management system purchased, coded and mapped to NPSA requirements • Trial planned of new incident reporting form Jan 05 • Investigation tools for all staff in production	• Further and wider spread incident reporting. • Investigations undertaken for all incidents relevant to incident grading and lessons shared across organisation.

Key area	Action	Lead	Timescale	Progress	Sources of assurance
Risk Register	• Awareness raising for all staff • Step by step guide • Awareness of assessment tools – link to tools already in place within the Health, Safety and Risk Control Workbook	Risk Coordinator All Managers	Ongoing	Step by step guide consulted upon and in production	Increased awareness and reporting of risks from all staff groups and via all reporting routes
Risk Management	Risk management training to be sourced	Risk Coordinator	On going	Discussions held with Deputy Director of Governance over the provision of training for all levels of staff	All staff have familiarity and understanding of risk management principles

References

Anderson P (2004). What is the impact of risk management in NHS Trusts? *Health Care Risk Report*, 10 (3) p12-13.

Clarke A and Dawson R (1999). *Evaluation Research*. Sage Publications, London.

Dickson G C A (2003). *Risk Analysis*. (3rd Edition). Witherbys, London.

Department of Health (1997). *The New NHS Modern and Dependable*. The Stationery Office, London.

Department of Health (1999). *Governance in the new NHS. Controls Assurance Statements 1999/2000: Risk Management and Organisational Controls*. HSC 1999/123. Department of Health, London. www.dh.gov.uk

Department of Health (2004). *The NHS Improvement Plan, Putting people at the Heart of Public Services*. The Stationery Office, London.

Gale F (2002). Challenges for Risk Managers. *The Healthcare Risk Resource*, 4 (2), 15-18.

HM Treasury (2001). *Management of Risk, A strategic overview, (The Orange Book)*. London.

Hopkin P (2002). *Holistic Risk Management in Practice*. Witherbys, London.

Institute of Chartered Accountants in England and Wales (1999). *Internal Control - Guidance for Directors on the Combined Code (the Turnbull Report)*. www.icaew.co.uk/internalcontrol

Jeyes J (2002). *Risk Management 10 Principles*. Butterworth-Heinemann, Oxford,

National Patient Safety Agency (2004). *Seven Steps to Patient Safety*. www.npsa.nhs.uk

Newark and Sherwood PCT (2003a). *Annual Report, Clinical Governance Report and Summary Financial Statements 2003/2004*.

Newark and Sherwood PCT (2003b). *Process for Adoption and Ratification of Policies and Procedures. IG012*.

NHS Litigation Authority (2004). *Risk Management Standard for Primary Care Trusts.* www.nhsla.com

Office of Government Commerce (2002). *Management of risk: Guidance for Practitioners.* The Stationery Office, London.

Standards Australia (2004). *Risk Management.* AZ/NZS 4360:2004. www.riskmanagement.com.au

Walshe K (2003). Understanding and Learning from organisational failure. *Quality and Safety in Healthcare* 12, 81-82

Waring A and Glendon A I (1998). *Managing Risk.* Thomas Learning, London.

West E (2001). Management matters: the link between hospital organisation and quality of patient care. *Quality in Health Care* 10, 40-48.

This page is intentionally blank.

5

Reducing claims against the NHS through the rapid and sensitive handling of complaints - I

ROBERT CALDEIRA

Introduction

The aim of this chapter is to establish whether the rapid and sensitive handling of complaints results in a reduction in the number of civil actions in law.

Wilson (1994) states that "to decrease the number of complaints proceeding to litigation, prompt handling of these complaints is required." Responding to patients' needs and wishes quickly and effectively has been the goal of health policy in the United Kingdom for several decades (Coulter 2002). The National Health Service Litigation Authority (NHSLA) supports this principle and encourages National Health Service (NHS) Trusts to provide explanations and apologies to patients (NHSLA, 2002). The General Medical Council (GMC) has also adopted this approach by requiring clinicians to inform the patient or family if, following an investigation, they discover that something had gone wrong (Capstick, 2004). The evidence to-date suggests that what patients want is to be listened to and to enter into meaningful dialogue with the medical practitioner. If the communication process breaks down, then this will potentially result in a complaint as patients see the complaints process as an avenue to getting their questions answered (Beckman et al, 1994; Eastaugh, 2004; Entman et al, 1994; Hickson et al, 1994; Levinson, 1994; Tingle, 1994; Vincent, 1994).

The mechanism for making complaints about one's experience within the NHS was first introduced in 1966 (Department of Health, 2003). In 1981 the then Department of Health and Social Security (DHSS) introduced guidance on making a complaint in Health Circular HC(81)5 (Miller 1986), although the system that was put in place was non-statutory. It took until 1985 to introduce a statutory complaints process, with Michael McNair-Wilson, a Conservative Member of Parliament, introducing a bill to establish a statutory procedure for dealing with hospital complaints (Miller 1986). The reason for the introduction of the bill was because Wilson had been the victim of a drug error, which had prolonged his stay in hospital. Instead of seeking financial redress, he introduced the bill to enact a statutory mechanism providing patients with information and a forum for their complaints that would allow the NHS to learn from adverse events (Miller 1986).

Mr Wilson's belief that it is important to find out *why* something went wrong rather than pursuing financial compensation (Miller 1986) is supported by the UK charity Action Against Medical Accidents (www.avma.org.uk/index.asp) who surveyed 2000 of their cases and found that approximately 70% of clients "are interested in financial compensation only as a secondary matter, or not at all. What they seek is an honest explanation of what went wrong and why. An apology if that is appropriate and an assurance that steps will be taken to ensure that an accident of that kind does not happen to anyone else" (Cited by Miller, 1986).

This belief is the basis of the complaints system today. In his "Making Amends" Consultation paper, the Chief Medical Officer says "The individual who has suffered harm as a result of the health care they have received must get an apology, clear explanation of what went wrong, treatment and care and where appropriate financial compensation. The NHS must also ensure that such bad experiences of individuals are learned from, so that future NHS patients throughout the country benefit from reduced risks and safer care. The primary aim must be to reduce the number of medical errors that occur" (Department of Health, 2003).

When someone makes a complaint in the NHS it is usually because they are dissatisfied with the way they or the person they are representing have been or are being treated and there is a breakdown in communication between the patient and the medical practitioner (Kiran and Jayawickrama, 2002). A number of research projects have concluded that complaints are often used as a means of obtaining information (Yamey 1999, Vincent, Young et al 1994) and as an attempt by the complainant or their representative to ensure that there is a learning outcome and that mistakes are not repeated (Anderson, Allan et al, 2000; Department of Health, 2003).

A number of studies have been undertaken to identify what leads patients to bring a medical negligence claim. Vincent et al (1994) undertook a survey of 227 patients and relatives to ascertain why they were taking legal action. The cohort was from five firms of solicitors. Over 70% of the respondents indicated that the decision to take legal action was not based solely on the original injury, but also on insensitive handling and poor communication. A two-part questionnaire was distributed which asked for the following information in the first part:

- A description of the incident which led to them consulting a solicitor;
- Emotional reactions to the incident;

- Effect on the patient's or relative's life;
- The quality of explanation received;
- Reasons for taking legal action, and
- The kind of help they most needed now

The second part of the questionnaire consisted of standard psychological questions relating to mood and emotional distress.

The following four main themes emerged from the analysis of reasons for litigation:

- Concern with standards of care – wishing to ensure that a similar incident did not happen again;
- Need for an explanation – wanting an explanation and feeling ignored or neglected after the incident;
- Compensation – Wanting compensation and an admission of negligence; and
- Accountability – a wish to see staff called to account and disciplined.

A further question was posed to the participants: "Once the original incident had occurred could anything have been done which would have meant you did not feel the need to take legal action?" A total of 41% of participants replied 'yes' and gave the following reasons:

- Explanation & apology;
- Correction of mistake;
- Pay compensation;
- Correct treatment at the time;
- Admission of negligence;
- Investigation by drug company/hospital
- Disciplinary action
- If listened to and not treated as neurotic;
- Honesty.

A study by Entman et al (1994) examined the relationship between the malpractice claims history of Florida obstetricians and the quality of the clinical care they provided to patients five to ten years after the claims. Doctors were classified into one of four groups based on claims history:

- High Frequency;
- High pay (frequent claims and high payments);
- No claims; and
- All others (intermediate experiences)

Nurses and Doctors who were not aware of the participant's claims history were used to review charts and assess quality of obstetric care. There was no demonstrable connection between the quality of care provided and the participants' prior malpractice claims history. The results are consistent with other studies that have come to a similar conclusion (Levinson, 1994). What was identified from the study was that breakdowns in communication between patients and doctors are critical factors leading to complaints and then on to litigation.

Vincent et al (1994) and Localio et al (1991) in their studies have also concluded that the quality of medical care alone is a poor predictor of a malpractice claim. The studies have shown that communication is a principal determinant of patients' evaluation of their treatment.

Hickson et al (1994) undertook a study in which the relationship between obstetricians' prior malpractice claims history and the satisfaction of patients with their obstetric care was examined. Doctors were classified into the same groups as in the study undertaken by Entman et al. As part of the process, mothers of still born infants and infants that died, together with a random sample of mothers with viable infants from the 1987 Florida Vital Statistics, were interviewed to assess their satisfaction with their obstetric care. The results indicated that patients of high frequency claims doctors felt rushed, feeling ignored, receiving inadequate explanations or advice and spending less time during routine visits with their doctors.

Beckman et al (1994) carried out a similar study and arrived at similar findings. They examined depositions in malpractice cases and identified four types of communication problems:

- Deserting the patient;
- Devaluing patients' views;
- Delivering information poorly; and
- Failing to understand patients' perspectives.

The study also concluded that these problems were present in more than 70% of malpractice depositions.

Levinson et al (1997), through their study of Physician-Patient communication, concluded that physicians who exhibit more negative communication behaviours are more likely to have been sued in the past for malpractice than those with more positive communication with the patient.

The Department of Health's own research shows that even when compensation is received claimants still want an apology and explanation (Cited in Department of Health, 2003)

It is difficult to obtain good information to show trends in claims prior to the NHS Litigation Authority taking over claims handling for the NHS in 1995. This is because, prior to 1995, Medical Defence Organisations handled claims and met any costs for primary care, and individual NHS Trusts handled their own cases (Department of Health, 2003).

However, an analysis by the medical defence organisations of complaints received from 1990 to 2000 found that the proportion going on to becoming claims was between two and six percent (Department of Health, 2003). This may suggest that the complaints were handled satisfactorily in line with NHS guidance (see table 1 below). The NHSLA says that 60 –70% of claims do not go beyond initial contact with a solicitor or disclosure of medical records (Department of Health, 2003). Those that do go on to become full claims may do so because they feel that all their questions have not been answered (Vincent et al, 1994).

Using information from the Medical Defence Union, Pearson (1978) showed that the number of claims in 1971 was double the average for the first 25 years of the NHS. Pearson states that "although claims were rising, averaging 1,000 a year against doctors, dentists, pharmacists or health authorities, they were still insignificant when compared with the six million in-patients treated each year." Information published by the Medical Protection Society and Medical Defence Union suggests that, between 1983 and 1987, claims doubled from 1,000 to 2,000 (Department of Health, 2003).

Since the NHS Litigation Authority took over handling NHS claims in the mid-nineties, the records they have kept from 1996-97 to-date are regarded as accurate (Department of Health, 2003). Figure 5.1 shows the total number of clinical negligence claims by financial year of incident as at 31 March 2004, including "below excess" claims handled by trusts. The information before 1996-97 may not be totally accurate because before this date the information was not centralised (Department of Health, 2003). The big increase in claims from 1996-97, compared to previous years, can be accounted for by more accurate record keeping by the NHSLA.

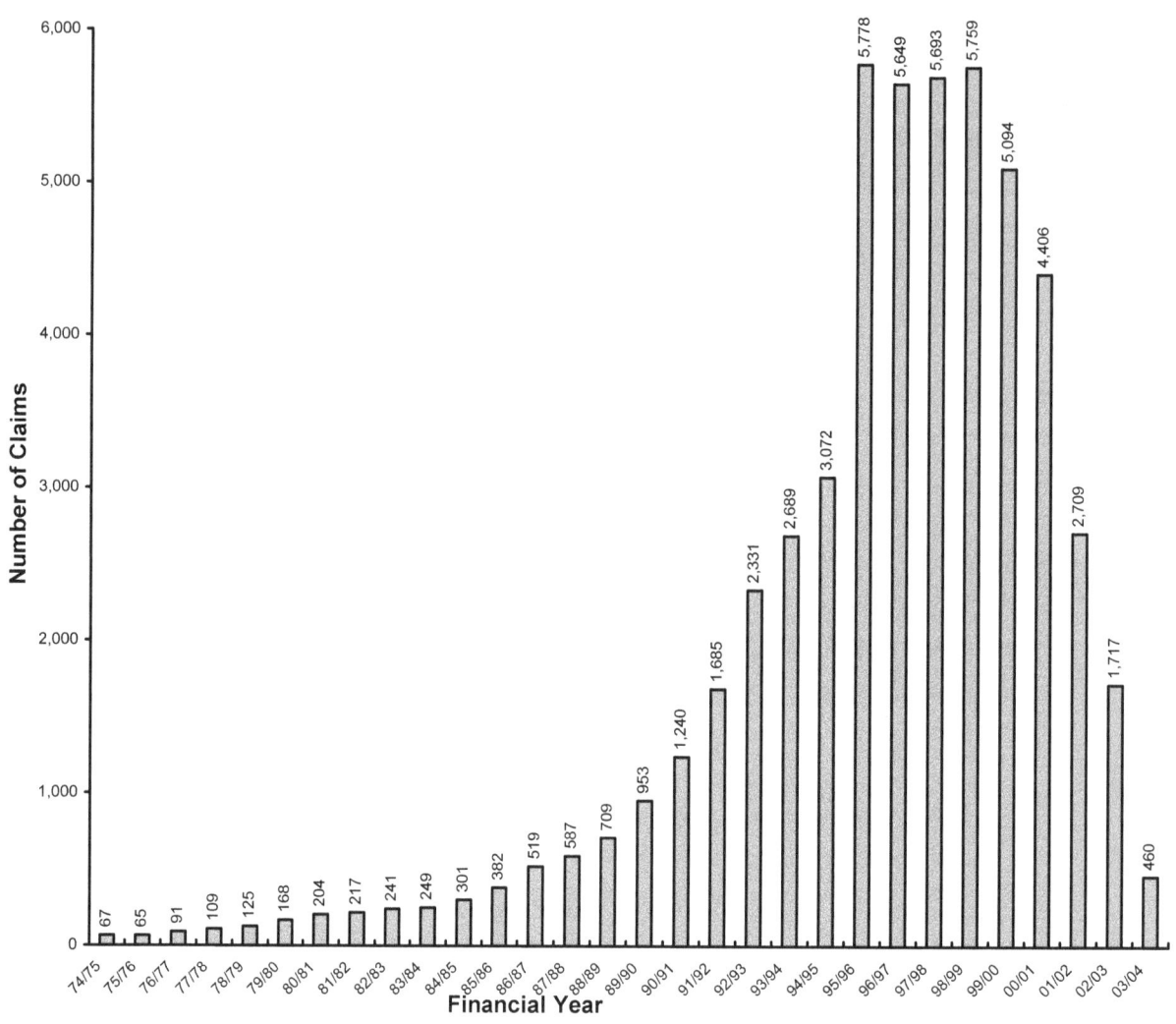

Figure 5.1 – Clinical negligence claims by financial year of incident

Table 5.1 - Written complaints about Hospital and Community Health Services: performance against targets for Local Resolution and Independent Review, England, 1996-97 to 2000-01 and 2002-03 & 2003-04

		2003-04	2002-03	2000-01	1999-00	1998-99	1997-98	1996-97
Total Written complaints received	No.	90,122	91,023	95,994	86,536	86,013	88,757	92,974
Local resolution:								
Local resolution concluded within performance target	No.	62,250	57,211	53,365	51,597	53,797	58,002	62,007
Local resolution concluded outside performance target	No.	23,465	28,646	36,396	30,120	28,309	27,210	26,992
Local resolution still being pursued at end of year	No.	4171	5,166	6,233	4,819	3,907	3,545	3,975
% of Claims Concluded At Local Level		95%	94%	94%	94%	95%	96%	96%
Requests for independent review:								
Cases requesting independent review	No.	2491	2,380	2,243	2,061	1,838	1,871	1,612
Cases still being considered	No.	661	582	446	394	336	370	283
Cases referred to independent review panel	No.	210	255	312	296	285	348	373
Independent review:								
Cases referred to independent review panel	No.	210	255	312	296	285	348	373
Independent review concluded within performance target	No.	83	50	72	80	64	88	131
Independent review concluded outside performance target	No.	113	72	120	99	112	96	67
Independent review still being pursued at end of year	No.	150	133	120	117	109	164	175

Figures for 1996-97 to 2000-01(Department of Health)
Figures for 2002-03 to 2003-04 (Department of Health)

2000-01 saw the introduction of the first Primary Care Trusts (PCTs). Figures for 2000-01 include information about written complaints received by the 40 PCTs which became operational during the year. 2001-02 figures have not been published because, according to the Department of Health, the quality of the data was suspect.

Table 5.1 shows that, for the years given, without exception around 95% of complaints were resolved at a local level.

If we look at the total number of complaints against the total number of claims by financial year we can see from the table 5.2, below, that the claims figures are in the range two to seven per cent of complaints figures, with the exception of 2003/04 [*Editor's note* – there is probably a simple explanation for 2003/04's anomalous result: insufficient time had elapsed for complaints to mature into claims]. These figures are in line with the Medical Protection Society and Medical Defence Union's analysis of the percentage of complaints that went on to become claims (Department of Health, 2003).

Year	1996/7	1997/98	1998/99	1999/00	2000/01	2002/03	2003/04
Complaints	92974	88757	86013	86536	95994	91023	90122
Claims	5649	5693	5759	5094	4406	1717	460
Percentage	6	6	7	6	5	2	0.5

Table 5.2 – Complaints and claims in the NHS by financial year

Concluding comments

Since the founding of the NHS in 1948, the public were initially slow in pursuing claims for medical/clinical negligence. This was due to their loyalty to the NHS and their relationship with the doctor (Miller, 1986). This relationship, however, began to weaken s due to the public's disillusionment with public services during the 1970's and the legitimisation of complaints through citizens' charters introduced in 1979 by the Conservative Government to hold public bodies to account (Hughes, 1999; Tingle, 1994). There has also been a marked change in social attitudes (Tingle, 1994). Patients and their representatives are no longer prepared to accept the status quo and now feel empowered to question the practitioner about their condition or treatment (Tingle, 1994). The change in social attitudes can be also attributed to a rise in public expectations, which has come about through access to education and higher attainment, which in turn has empowered people to question what affects them, in this instance health matters (Hughes, 1999).

The evidence shows that the road to litigation begins with the breakdown in communication between the practitioner and the patient (Eastaugh, 2004; Beckman et al, 1994; Entman et al 1994). This may lead to a complaint being pursued, as this is seen as a way of eliciting information (Yamey, 1999; Vincent et al, 1994) or the legal route which may provide compensation and sometimes an explanation but does not really address the patient's concerns about standards of care or accountability (Vincent et al, 1994). The Department of Health's own research has shown that claimants who received compensation still wanted an explanation and an apology (Department of Health, 2003)

The available statistics (Department of Health, 2005a; Department of Health, 2005b) suggest that if complainants' issues can be addressed locally, and we have seen that currently around 95% of complaints are resolved, then this reduces the number that will go on to litigation. This is supported by the Medical Protection Society's analysis from 1990–2000 of complaints and the percentage of those that went on to become claims: over the decade it was between two and six per cent (Department of Health, 2003).

When we examine the statistics, however, we can see that complaints are in the region of 90,000 a year and claims are around six per cent of complaints levels (Department of Health, 2005a; Department of Health, 2005b). Furthermore, on average 850,000 adverse incidents occur within the NHS annually (Department of Health, 2003) and of those only 10% result in complaints (Department of Health, 2005a; Department of Health, 2005b). This would suggest that some intervention is taking place to stop a complaint being made. In part, this may be due to communication in an open and positive manner with the patient, as advocated by the General Medical Council (Capstick, 2004) and by the NHS Litigation Authority (2002).

The USA and Canada recognise the importance of effective communication between doctors and patients. From 2005 the USA is reintroducing a requirement for medical school graduates to pass a national test for communication and clinical skills before they can obtain a license to practise; this test had been previously dropped in 1964. The Medical Council of Canada have a similar test (Eastaugh, 2004).

In looking at the evidence, communication plays an important role in stopping complaints and claims, and in stopping complaints progressing to claims. In the event of a complaint being made, the statistical evidence appears to support the hypothesis that rapid and sensitive handling of complaints does indeed result in a reduction in claims.

Editor's note

Readers might like to note that the issue of clinical negligence claims in the NHS has been considered in some depth in a report by the National Audit Office *Handling Clinical Negligence Claims in England* (HC 403 of 2000-2001, published on 3 May 2001), which can be viewed at, and downloaded from www.nao.org.uk .

The key findings in the report were:

(a) At the time the report was published, the number and value of claims was increasing. At March 2000, there were 23,000 claims outstanding, with a net present value of £2.6 billion. And claims expected to arise from incidents that may have occurred but not been reported are valued at a further £1.3 billion.
(b) On average, claims took a long time to settle.
(c) Nearly half of the claims settled in 1999-2000 cost more in legal and other costs than the settlement itself. For settlements up to £50,000, the costs of reaching the settlement were greater than damages awarded in over 65 per cent of cases.
(d) The initiatives that had been taken by the Legal Services Commission and the NHS Litigation Authority to improve the quality of solicitors advising on bringing and defending clinical negligence claims were having a positive impact.

And the main recommendations were:

(a) The Litigation Authority should draw up an action plan with quantified targets and performance measures to address claims that have been open for more than five years. The Legal Services Commission should, similarly, monitor the progress of cases over five years old, and take steps to bring them to a timely conclusion. The Litigation Authority and the Legal Services Commission should hold regular meetings to consider general concerns in concluding cases but not, of course, specific cases.

(b) The Department of Health, Lord Chancellor's Department and the Legal Services Commission should further investigate alternative ways of satisfactorily resolving small and medium sized claims, for example through the offering of the wider range of non-financial remedies that patients say they want, setting up regional panels and offering mediation where appropriate.

References

Anderson K, Allan D, Finucane P (2000). Complaints Concerning the Hospital Care of Elderly Patients: A 12-Month Study of One Hospital's Experience *Age and Ageing* September 29(5) 409.

Beckman H, Markakis M (1994). The Doctor Patient Relationship and Malpractice Lessons from Plaintiff Depositions. *Arch Intern Med* Vol 154 1365-1370.

Capstick B (2004). The future of clinical negligence litigation, *BMJ* 328 457–459

Coulter A (2002). After Bristol: Putting Patients at the Centre. *BMJ* 324 648-651.

Department of Health. Complaints data. www.dh.gov.uk

Department of Health (2003). *Making Amends*. Department of Health, London

Eastaugh S (2004). Reducing Litigation Costs Through Better Patient Communication. *The Physician Executive* May-June p. 36-38.

Entman S, Glass C (1994). The Relationship Between Malpractice Claims History and Subsequent Obstetric Care. *JAMA* Vol 272(20) 1588-1591.

Hughes C (1999). *European Journal of Health Law.* 6 p. 45-53.

Kiran T, Jayawickrama N (2002). Complaints and Claims in the UK National Health Service. *Journal of Evaluation in Clinical Practice* Vol 8 (1) 85-86.

Levinson W (1994). Physician-Patient Communication: A key to Malpractice Prevention. *JAMA* Vol 272(20) 1619-1620.

Levinson W (1997). Physician-patient Communication: The Relationship with Malpractice Claims among Primary Care Physicians and Surgeons. *JAMA* Vol 277 553-559.

Localio A R, Lawthers A G, Brennan T A, Laird N M, Hebert L E, Peterson L M, Newhouse J P, Weiler P C, Hiatt H H (1991). Relation between malpractice claims and adverse events due to negligence: Results of the Harvard Medical Practice Study III. *NEJM* 1991; 325: 245-251.

Miller F H (1986). Medical Malpractice Litigation: Do the British Have a Better Remedy? *American Journal of Law & Medicine.* 11(4): 433-63.

NHS Litigation Authority (2004). *Fact Sheet 3: Information on Claims.* www.nhsla.com

NHS Litigation Authority (2002). *Apologies and Explanations Circular No 02/02.* www.nhsla.com

Pearson C (1978). *The Report of the Royal Commission on Civil Liability and Compensation for Personal Injury.* HMSO, London.

Tingle J (1994). Adopting Strategies to Reduce Healthcare Litigation and Complaints. *British Journal of Nursing* Vol 4 No. 7 p. 17.

Vincent C, Young M, Phillips A (1994). Why do People Sue Doctors? A study of Patients and Relatives taking Legal Action. *Lancet* 343 1609-13.

Wilson A (1994). *Being Heard. The Report of the Review Committee on NHS Complaints Procedures.* Department of Health, London.

Yamey G (1999). Report Condemns National Health Service Complaints Procedure. *BMJ* 319 p. 804.

6

Reducing claims against the NHS through the rapid and sensitive handling of complaints - II

JAYNE HARTLEY

Introduction

The intention of this chapter is, building on chapter 5, to further discuss whether the rapid and sensitive handling of complaints made by recipients of care and treatment in the National Health Service (NHS) can lead to a reduction in the number of civil actions in law.

Before commencing this discussion, the key elements of the complaints process within the NHS will be explored. This will include an understanding of key terminology, the development of the present complaints process and the role of significant people and organisations, including the Patient Advice and Liaison Services (PALS), the Healthcare Commission and the NHS Ombudsman. The claims management process in the NHS will then be described. This will address the specific terms and procedures that relate to claims management, and the civil law relating to claims for compensation for clinical negligence.

I will then explore in more detail the incidence and costs of complaints and claims that currently exist within the NHS. This exploration will include reference to national data collated by the Department of Health (Department of Health) and the NHS Litigation Authority (NHSLA). The factors which may influence the number of civil actions in law will then be considered. This will include the management of complaints and the impact of the complaints process that currently exists within the NHS. However, other issues will also be addressed which may affect the complainants' decision making processes, for example: appropriate communication with healthcare personnel and the provision of sufficient information to make an informed choice. The effects of other factors within an NHS organisation which may influence the decision to take legal action will also be considered. This will focus on clinical governance and risk management systems, including adverse incident and near miss reporting processes and staff training. The impact of factors external to an NHS organisation will also be explored, including pre action protocols, alternative dispute resolution (ADR) and conditional fee arrangements.

The complaints process in the NHS in England

A complaint is described by Debell (1997) as "an expression of dissatisfaction", whilst Longman (1984) describes a complaint both as "an expression of discontent" **and** a "formal allegation by a plaintiff in a civil action" which forms a direct link between complaints and claims. Interestingly, the NHS complaints regulations (Department of Health, 2004) do not clarify the definition of a complaint – despite an extensive glossary.

Over recent years, the Department of Health has introduced two fundamental reforms to the complaints process – the first and most radical in 1996 and the second in 2004. These will now be discussed.

In the early 1990's pressure from consumer groups and healthcare staff led to the development of a committee chaired by Lord Wilson to review the complaints processes in the NHS (Mulcahy, 1999). The findings of the Wilson Committee were published in May 1994 and reported that existing arrangements for handling complaints were "too complex, too lengthy and too confrontational" (Department of Health, 1994). Radical changes were recommended and were virtually all adopted by the Government and brought into force in 1996 (Shipman Inquiry, 2005).

The 1996 guidance on the implementation of the NHS complaints procedure (Department of Health, 1996) stipulated that each NHS organisation should establish and publicise a complaints procedure which consisted of three key stages (Mayberry 2002). The first required implementation of a process of *local resolution*, which involved direct communication between the complainant and the Trust, either by letter or face to face in an attempt to resolve the complaint. Local resolution remains the first stage of the revised 2004 complaints process, although the updated guidance requires the presence of a senior member of staff at this stage (Department of Health, 2004).

The second stage necessitated referral to an *independent review panel* if the complainant remained dissatisfied following attempts at local resolution. However, this was not an automatic part of the process. The decision to set up such a panel lay with a convenor – who was usually a non executive director of the NHS organisation involved in the complaint (Roberts, 2002 p277). The independent review panel consisted of three members: the convenor, an independent lay chairman chosen by the NHS regional office and a lay representative of the purchasing health authority (Department of Health, 1996). Its purpose was to investigate the complaint, draw conclusions and make suggestions (Mayberry, 2002). One of the most significant changes instigated as a result of the 2004 complaints process has been to transfer the responsibility of the independent review panel to the Healthcare Commission. This is an independent body established to promote improvements in healthcare (www.healthcarecommission.org.uk), which will either make recommendations to the Trust concerned as to what action should be taken to resolve the complaint, or convene a panel to deal with the complaint. The panel consists of three people, but "any person who is a member or an employee of an NHS body or a person who is, or has been, a healthcare professional or employed by a healthcare professional will not be a member of the panel" (Department of Health, 2004). This is in significant contrast to the independent review panel that had NHS employees as its members.

The third stage of the process continues as per the 1996 complaints procedure (Department of Health, 1996) and allows the complainant, if they remain dissatisfied with the outcome of the investigations that have taken place, to ask the Health Service Commissioner or Ombudsman to investigate. The Healthcare Commission can also refer cases to the Ombudsman, who is also independent from the NHS.

The Ombudsman will investigate complaints relating to clinical care and administration matters that have occurred in the NHS and will either uphold the complaint in part or in full, or may not uphold the complaint at all. The Ombudsman may make recommendations and, if so, will expect that these are implemented (Roberts 2002, p281). Twice yearly reports are produced which summarise the outcomes of selected investigations and identify areas of poor practice (Health Service Ombudsman, 2004).

The complaints procedure (Department of Health, 1996) also required that a designated person should be identified to deal with complaints as they arose. Some trusts chose to appoint a complaints manager, whilst others placed this responsibility on existing post holders. It was even suggested within the 1996 Department of Health guidance that the Chief Executive could be the complaints manager – although this has been removed from the 2004 complaints procedure, which requires that the role of the complaints manager be strengthened and their profile raised (Department of Health, 2004). The 1996 complaints guidance (Department of Health, 1996) discussed the role of Community Health Council (CHC) staff as providing support for complainants during the complaints process. The community health councils were seen as "the voice of NHS consumers" and much concern was expressed following the decision to abolish them as part of the NHS Plan in 2000 (Eaton, 2003).

The NHS Plan (Department of Health 2000) announced that CHCs would be replaced by a Patient Advocacy and Liaison Service (PALS) which would be available in every trust by April 2002. The PALS officer is an employee of the trust but is required to remain impartial and represent the patients of that organisation by providing confidential advice and support for patients, families and carers. They also have a fundamental role in attempting to resolve problems and concerns quickly and providing information about the NHS complaints procedure.

The importance of the PALS role, together with the necessity to train those staff involved in the complaints process, is reinforced in the 2004 complaints guidance (Department of Health, 2004). Interestingly, there has been no change to the requirement that the complaints process must cease if the complainant makes it explicit that they intend to take legal action in respect of the complaint (Department of Health, 2004). This conflicts with the recommendation made in 'Making Amends' (Department of Health, 2003) where it is stated that: 'the rule in the current NHS complaints procedures requiring a complaint to be halted pending resolution of a claim should be removed as part of the reform of the complaints procedure'. The Chief Medical Officer, who wrote this report, felt this change might reduce the number of people who pursued a formal litigation process – and might also lessen the dissatisfaction complainants and claimants felt at the end of the process (Department of Health, 2003).

The claims management process in the NHS in England

In England, compensation for medical harm is largely dealt with through a legal process that begins with an allegation of medical negligence known as 'tort' law (Department of Health, 2003). Tort law is an area of civil law. A 'tort' is an act or omission by the defendant that causes damage to a claimant's property, reputation or interests (Cooke, 1997 p3). It can be demonstrated as follows:

Act (or omission) + Causation + Fault + Damage = **Liability**

(Adapted from Cooke, 1997 p4).

Negligence is the most important tort in modern law (Elliott and Quinn, 2002 p477) and the commonest tort claim (Adams 2003, p140). It may be defined from a clinical perspective as "a breach of duty or care that directly results in injury or loss" (Chapman, 2001 p541). To be successful in a claim of negligence, under the law of tort, a patient who suffers harm must prove on the balance of probability that the hospital which owes the duty of care has caused the injury or damage through care delivered in a negligent manner (Department of Health, 2003).

Civil law differs greatly from that of criminal law, and the key differences are summarised in Figure 6.1, below.

Criminal Law	*Civil Law*
Provides the machinery by which the state may take action against offenders.	Gives legal rights to individuals to enforce the rights governed by relationships.
Proceedings generally started by the police or local authorities – the victim usually plays no part in the decision to prosecute.	Proceedings started by the individual who believes they have been wronged.
The case will proceed to trial in a magistrates court or crown court.	Most civil cases are heard in the county court or high court.
Prosecution must prove guilt **beyond all reasonable doubt.**	The claimant must prove the defendant is liable on the **balance of probability**.
Punishment is the outcome of criminal legal action and may include imprisonment.	Damages and compensation may be the outcome of civil legal action.

Figure 6.1 - A summary of key differences between criminal and civil law (Adapted from Adams, 2003, pp 5-6).

To manage the claims process in the NHS, each trust is required to have a named person with responsibility for the management of claims. In some trusts this role is combined with other responsibilities, which might include complaints management. It is the claims manager who will, most often, be the first point of contact for the claimant's solicitor, and to be notified that a claim is pending. The claims management process may be summarised as follows (Chapman, 2001 pp 547–550):

- **Request for records** - the claims manager must arrange for copy records to be provided within 40 days.
- **Initial internal investigation** – this will be led by the claims manager, although with the advent of more robust incident reporting systems, it is possible that this will have already taken place.
- **Letter of claim** – this is issued at least three months prior to the instigation of formal legal proceedings and details the nature of the alleged negligence, resultant damage and a valuation of the claim. This process is part of a pre action protocol for clinical disputes which attempts to settle the claim at this point before going to court (Clinical Disputes Forum, 2002).
- **Letter of response** – The defendant trust has 14 days to acknowledge the letter of claim. It then has three months to prepare a report detailing the

summary of the case, the breach of duty, causation, quantum and the future plan to manage the claim – for example to defend or settle. This is a relatively short time scale to ensure the claims process continues to moves forward and does not falter.

- **Issue of proceedings** - If resolution has not been achieved by this stage, then the formal litigation process begins.
- **Statement of claim** – This is issued once it has been agreed that legal proceedings are to take place. This must be acknowledged by the defending trust within 14 days, and a defence sent to the court within a further 14 days.
- **Case management** – The court now sets the timetable for what will happen next, including the trial date if an out of court settlement cannot be agreed. A trial can be held to determine both liability (whether negligence has occurred) and quantum (how much compensation is to be awarded).

Responsibility for the management of clinical negligence claims made against NHS bodies in England lies with the NHS Litigation Authority. The NHSLA was established in 1995 as a special health authority and is charged with ensuring there is a fair outcome from their interventions for both the patient and the NHS (Department of Health, 2003). The NHSLA handles negligence claims under five separate schemes (NHSLA, 2004) including the Clinical Negligence Scheme for Trusts (CNST). This is a risk pooling scheme for clinical negligence claims funded from members' contributions. Although membership is voluntary, currently all NHS trusts and Primary Care Trusts (PCTs) in England choose to belong (NHSLA, 2004).

The incidence and cost of complaints and claims in the NHS

The Department of Health provides annual statistics relating to the number of written complaints received about hospitals and community health services in England (these may be accessed via the Department of Health webpage on www.dh.gov.uk) and are shown in Figures 6.2a and 6.2b.

YEAR	96/97	97/98	98/99	99/00	00/01	01/02	02/03	03/04
TOTAL	92,974	88,757	86,013	86,536	95,994	*	91,023	90,122

Figure 6.2a

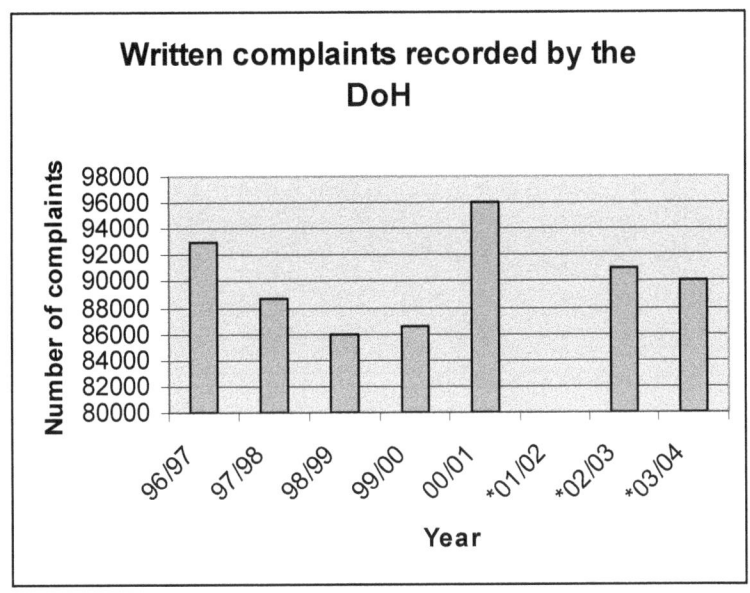

Figure 6.2b

*According to the Department of Health, there was an error in the data collection which meant it could not be validated – therefore these statistics are not available.

The Department of Health complaints reports also provide annual figures relating to the number of written complaints where an independent review was requested, and those which were actually referred for independent review by the convenor. These are shown in Figures 6.3a and 6.3b.

YEAR	96/97	97/98	98/99	99/00	00/01	01/02	02/03	03/04
Independent reviews requested	1,612	1,871	1,838	2,061	2,243	*	2380	2491
Referred for independent review	373	348	285	296	312	*	255	210
Variance	*1,239*	*1,837*	*1,553*	*1,765*	*1,931*		*2,125*	*2,281*

Figure 6.3a

83

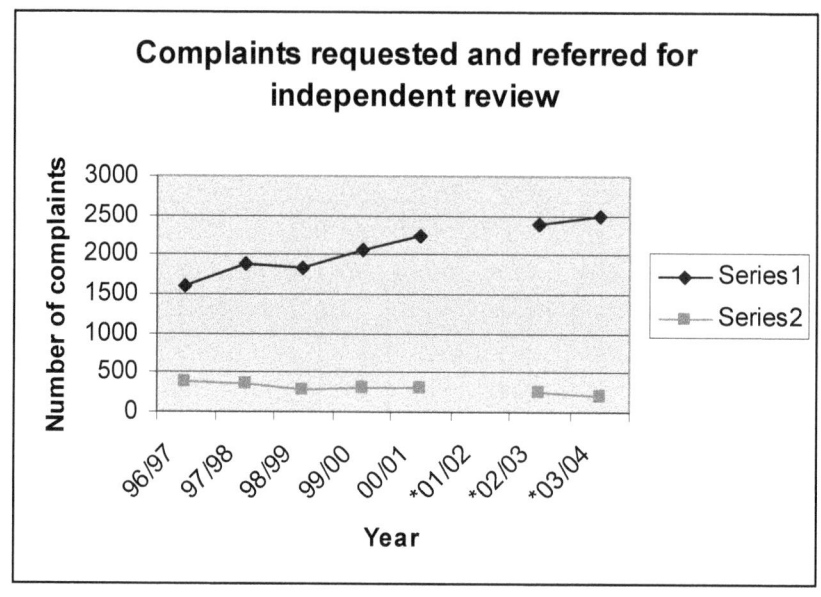

Figure 6.3b

On Figure 7.3b, 'Series 1' relates to those complaints where an independent review was requested, whilst 'Series 2' relates to those that were actually referred for independent review.

The Ombudsman also maintains a record of the number of number of complaints received, and these are recorded in Figure 6.4.

YEAR	Number of Complaints received
1996/1997	2,219
1997/1998	2,315
1998/1999	2,511
1999/2000	2,526
2000/2001	2,595
2001/2002	2,651
2002/2003	3,994

Figure 6.4

In Figure 6.5 it can be seen that, whilst the number of complaints received by the Department of Health has fluctuated over the years since 1996/1997, complaints to the Ombudsman have continued to rise. This could possibly reflect increasing dissatisfaction with the earlier stages of the complaints process.

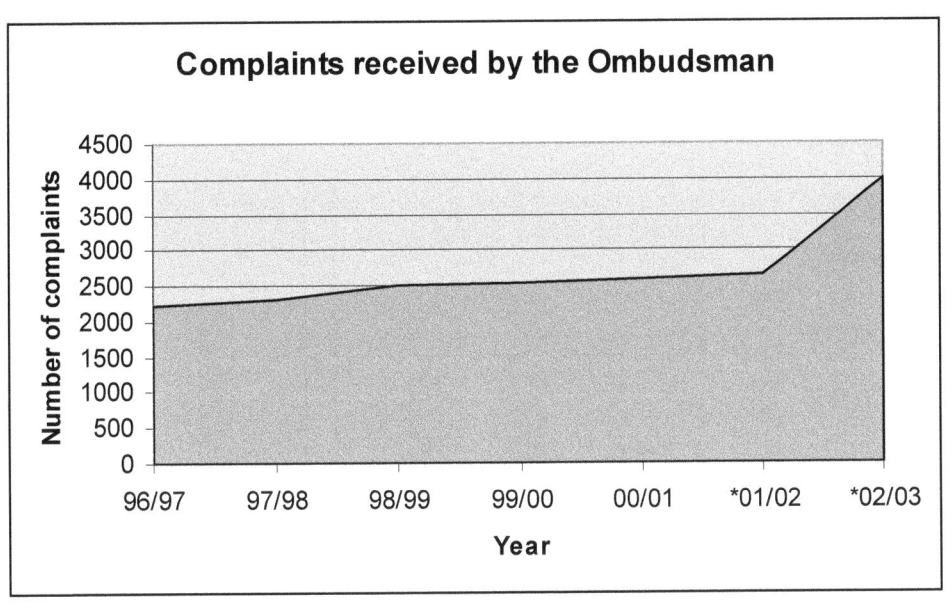

Figure 6.5 – NB: The asterisks on 01/02 are not explained.

There are also costs associated with the effective management of complaints. In 1995, the Cabinet Office Complaints Task Force estimated that the costs of dealing with a single complaint internally was approximately £410. These costs increased to £770 if an independent review was required and £11,200 if the Health Service Commissioner became involved. Although these figures are now 10 years old, it can be seen that dealing with complaints is an expensive business.

The NHSLA has paid out varying and substantial sums of money in respect of negligence claims against the NHS since its inception in 1995 (NHSLA, 2004b). These are shown in Figures 6.6a and 6.6b.

YEAR	96/97	97/98	98/99	99/00	00/01	01/02	02/03	03/04
TOTAL (£'000)	6,015	49,460	70,076	277,746	872,966	553,986	467,530	432,644

Figure 6.6a

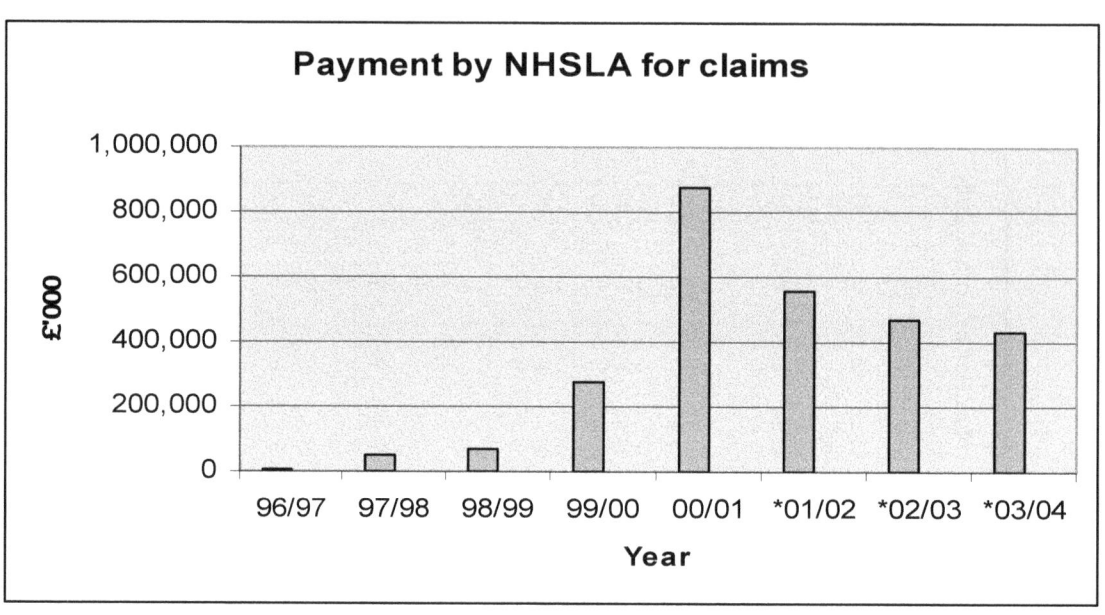

Figure 6.6b

It can be seen that there was a significant increase in the payment of litigation claims in 2000/2001 which has not been repeated since that time. The figures provided actually show a gradual annual reduction so that the data for 2003/2004 is the lowest level for four years. Further data from the NHSLA (2004b) show that in 2002–03, they received 7,798 claims of clinical negligence and 3,667 claims of non clinical negligence. In 2003–04, they received 1,547 fewer claims of clinical negligence (6,251) with 152 more claims of non clinical negligence (3,819). This means there was a total of 1,395 fewer claims made in 2003–04 compared with the previous year. As a result of these data, the NHSLA (2004a) states that 'although much is written in the press about the "compensation culture" ... this does not reflect our experience.' The possible reasons for these findings will now be discussed.

Factors influencing civil actions in Law

a) <u>Complaints management</u>

It has been shown that the introduction of the complaints process by the Department of Health in 1996 appears to have done little to reduce the incidence of complaints and litigation costs. Whilst the overall number of complaints recorded by the Department of Health has remained fairly constant, the number of complaints referred for independent review and to the Ombudsman have increased significantly. It might therefore be concluded that local arrangements for dealing with complaints at a local level have not been particularly effective, despite the introduction of the PALS officers. The reasons for this conjecture are not difficult to find. Complaints are often time-consuming to investigate and require the ability of the investigator to deal diplomatically with staff who may feel very defensive since "complaints frequently lead to the allocation of blame, disciplinary action or litigation ..... and there appears to be little reward for good complaint handling" (Allsopp and Mulcahy, 2001 p497).

A study undertaken by Jain and Ogden (1999) examined how general practitioners (GPs) felt about patients' complaints. They found that GPs experienced negative feelings of shock, anger, depression and had doubts about their clinical competence which were resolved by either practising defensively or planning to leave general practice, or - in some cases – were not resolved at all. These findings are replicated in the study undertaken by Cunnningham (2004), who analysed the impact on doctors in New Zealand of receiving a medical complaint. 221 doctors included in the survey indicated that they felt anger, depression, shame, guilt and persistent feelings of reduced goodwill and trust towards patients in general (not just the complainant).

Cunningham's (2004) survey also found that there was no evidence that the receipt of a complaint improves the delivery of patient care. Similarly, Jain and Ogden (1999) describe how only a small number found the complaint to be a learning experience. This is hardly positive publicity for complaints management. Complaint management by staff who have such negative feelings could lead to poor handling of the complaint and the complainant. This could mean that complaints would be viewed as low priority, undergo inadequate investigation and staff could have limited interaction with a complainant who could be viewed as troublesome and someone to be avoided. These behaviours are likely to encourage complainants to seek alternative methods of redress including civil legal action.

Consequently, it is important to acknowledge and understand the positive aspects of complaints so that they are handled appropriately. Complaints, like adverse incident reporting, can identify practices which, if repeated, could compromise the safety of other patients. They may therefore be used effectively as an early warning system. Complaints also present a valuable insight into how a service is perceived by the people who use it and also present an opportunity for engagement with service users (Allsopp and Mulcahy, 2001 p497).

If complaints can be dealt with in such a positive culture, then it is likely that the complaints will be dealt with in a timely manner with appropriate sensitivity so that local resolution may be effective. This will require staff involved in the complaints process to be appropriately trained and to ensure that the lessons learned from complaints are implemented and shared throughout the organisation. These actions should also have a positive effect upon the incidence of claims since the issue which led to the original complaint will have been resolved. It seems reasonable to assume that if situations that lead to complaints can be dealt with then that would have more effect on reducing legal action than the complaints process itself.

There is little research evidence to link the management of complaints with the incidence of claims, although Hickson et al (2002) undertook a retrospective study of the computerised records of 645 physicians and discovered that patient complaints were positively associated with the physicians' lawsuit experiences, so that those physicians who had received most complaints also had most lawsuits ongoing - even after adjustment for clinical activity. However, this study did not review how the complaints were managed, but simply recorded the numbers of complaints.

It will be interesting to discover whether the revised complaints process (Department of Health, 2004) will have an impact on the current trend of complaints. The introduction of a senior staff member into the local resolution process will undoubtedly raise the profile of the complaint in the eyes of the complainant and the trust – which could be construed as positive, in that the complainant might feel their complaint is being taken seriously and necessary changes will be implemented – or negative, in that the complainant could feel the trust is unusually concerned about their complaint and therefore they would have a better chance of success should they decide to go for litigation.

Local resolution intends to enable complainants to be dealt with promptly and at the point of service delivery (Allsopp and Mulcahy, 2001 p500). However, the increasing number of complaints referred to the Ombudsman could indicate dissatisfaction with the earlier stages of the complaints process and one possible solution could be to introduce an independent person into the local resolution stage, so that complainants feel their complaint is being investigated without bias. The PALS officer, although intended to undertake this role, may not be seen as independent, as they are employed by the organisation that is dealing with the complaint.

The impact of the Healthcare Commission remains to be seen, but it is debatable whether the process of independent review from the 1996 complaints system (Department of Health, 1996) adequately addressed the disquiet of the vast majority of people who did not progress beyond the local resolution stage. It has been shown that a significant number of requests for independent review (between 1,239 and 2,281) were turned down by the convenor; this must raise questions as to the partiality of the trust non-executive directors who held that position (Mayberry, 2002).

The review of the complaints process and the lack of research evidence does not confirm the impact of effective complaints management. But it is reasonable to assume that, if a complaint is well handled from the outset, with promptitude and sensitivity, it would be less likely that the complainant will consider involving lawyers or indeed referring the matter to the Health Commission (Mulvaney, 2004). However, there are undoubtedly other factors which could impact on the decision to make a claim for negligence against the NHS and these will now be considered.

b) Adverse incident and near miss reporting

The use of an integrated incident reporting system can provide an early warning system for potential complaints and claims. This enables healthcare providers to act proactively and promptly to deal with incidents, apologise, investigate and feed back the results and action to be taken before the patient has even thought about making a formal complaint.

It is of some concern that the Chief Medical Officer's report 'An Organisation with a Memory' (Department of Health, 2000b) suggests that there are as many as 850,000 serious adverse healthcare events occurring in the NHS hospital sector each year as a result of 'medical errors'. Reducing the incidence of adverse incidents requires careful analysis of every incident to determine both the multifactorial causes and good practices that can help minimise repetitions (Amoore and Ingram, 2002). Such action can only benefit patient care and reduce the possibilities for complaint and litigation (Shaw, 2004).

Hinckley (2003) agrees that making efforts to control adverse incidents can help to minimise litigation, whilst the reporting of near misses also offers numerous benefits such as greater frequency recording, which allows quantitative analysis and recovery patterns that can be captured, studied and used for improvement (Barach and Small, 2000).

c) Clinical governance

The introduction of clinical governance is also encouraging a greater focus on patient care and the development of early warning systems that highlight inadequate systems of care in advance of complaints and claims. There are seven pillars of clinical governance: clinical effectiveness, patient, carer and public involvement, staffing and staff management, education and training, use of information, clinical audit and risk management (Commission for Health Improvement, 2002), all of which have a significant impact on patient care. Golten (2000) sees clinical governance as having a crucial role in shaping the way healthcare providers respond to clinical negligence claims and thereby reducing the cost of litigation to the NHS. Most specifically, risk management is likely to have a critical impact on the incidence of litigation.

The establishment of the Clinical Negligence Scheme for Trusts (CNST) through the NHSLA is intended to "help raise standards of care in the NHS and hence reduce the number of incidents leading to claims" (NHSLA, 2004a). The CNST assessment process has ensured that certain risk management standards and processes are implemented in NHS organisations that restrict the possibilities of clinical negligence occurring. For example, systems must be established to ensure that staff only undertake that for which they have been trained and assessed as being competent to undertake.

Similarly, staff must show that they have received training in specific areas that are considered essential to promote patient safety, such as basic life support, blood product handling, principles of risk management and infection control. Clearly, any system as described above which is designed to reduce errors and adverse incidents should, if successfully implemented, also reduce the number of claims (Harpwood, 2001).

d) Communication

The publication 'Good Medical Practice' (General Medical Council, 2001) stresses that "good communication between patients and doctors is essential for effective care and relationships of trust". It requires doctors to give patients information about their condition, treatments and prognosis and to ensure that their informed consent has been obtained before any treatment is given. The guidance also requires that when a patient has suffered harm, the doctor should explain what has happened, act to put matters right where possible, and apologise.

Levinson et al (1997) undertook an audiotape analysis of 124 physicians to determine if specific communication styles could be linked to malpractice claims. Their findings showed that those physicians who used communication techniques such as checking patients' understanding, soliciting their opinions, educating patients about what to expect and encouraging them to talk had significantly fewer malpractice claims that those physicians who did not use these techniques. Whilst this may be a simplistic interpretation of the reason for lawsuits, the issue of communication is undoubtedly a factor in determining whether or not a claim may be made.

These findings are reinforced by Wilson (1998) and Gorney (1999), who both cite communication as the most powerful tool in clinical practice and believe that communication is not only vital for quality care but also for the avoidance of "litigious intentions amongst patients" (Wilson, 1998).

Studies into the impact of communication on litigation have continued. Moore et al (2000) undertook a study with 104 obstetric patients and discovered that "positive physician communication behaviours increased patients' perceptions of physician competence and decreased malpractice claim intentions towards both the physician and the hospital." These findings are similar to those of Krause et al (2001) who evaluated 178 medical expert opinions about complaints that had been made and discovered that a considerable proportion of lawsuits originated from misunderstandings rather than treatment errors; and, in 2002, Ambady et al discovered that a surgeon's tone of voice in routine visits was associated with malpractice claims.

e) Access to Justice

Reforms to the operation of tort law were proposed following an investigation by Lord Woolf into the civil law processes in England and Wales. The findings were reported in 'Access to Justice' (Woolf, 1996) and highlighted that many cases could be resolved without the need for court involvement if a greater spirit of co-operation could be established at the early stages of a potential claim.

To assist in this process the Clinical Disputes Forum (CDF) was formed in 1997 (Clinical Disputes Forum, 2002). It aimed to develop protocols which increased pre action contact between the claimant and the defendant, improved information exchange and investigation and put both sides in a better position to consider the settlement of cases fairly and promptly without recourse to litigation (Roberts, 2002 p286).

The CDF also intended to allow for settlements that included explanations of treatment, arrangements for follow up, letters of apology and sharing evidence of changes in practice, as well as resolving the claim itself from a financial perspective (Chapman, 2001 p545). A survey of the Association of Personal Injury Lawyers cited by the Department of Health (2003) showed that 33% of all cases avoided litigation as a result of the pre action protocol.

Lord Woolf also placed significant emphasis on 'Alternative Dispute Resolution' (ADR) as a means by which medical negligence claims could be resolved without going to trial. Mediation, or facilitated negotiation, is the most common form of ADR and since May 2000 the NHSLA has been requiring solicitors representing NHS trusts to consider every case for mediation (Department of Health, 2003). The CDF (2001) has issued a detailed guide to mediation in conjunction with the NHSLA. In this process, the mediator is a neutral third party, who may or may not be legally trained, who listens to evidence from both sides and then offers a solution which may include financial compensation (Osbourne, 2002, p456). The NHSLA reports that numbers of mediations rose from nine in 2000 to 47 in 2002, but these numbers are obviously low and approximately 65% of offers of mediation are refused by claimants (Department of Health, 2003). One reason for this might be the cost, since mediation is not cheap whilst it is also recognised that there is a lack of trained mediators available at short notice (Department of Health, 2003). Whilst mediation does not currently seem to be making a significant impact on the number of civil actions in law, it is still early days and may continue to evolve.

f) Conditional fee arrangements

In 1999, the Lord Chancellor's Access to Justice Act (based on the 1996 Access to Justice inquiry) introduced major reforms to the provision of state-funded legal services (Elliott and Quinn, 2002 p195). In essence, the reforms abolished public funding (legal aid) for most personal injury cases on the basis that other methods of funding – e.g. conditional fee arrangements – were available and more appropriate (Sandbach, 2004). This move was not without controversy, but the result is that 'no win no fee' funding is now the industry standard. Within this scheme, solicitors agree to charge no fees unless the case is won, although claimants may have to pay court and expert witness fees and take out insurance against losing and having to pay the other side's costs (Dyer, 2004).

In the past, patients may have been deterred from claiming because they could not afford to do so. However, the advent of conditional fee arrangements, the increase in the number of specialist solicitors dealing with medical negligence and the advertising of their services together with changes in patient attitudes toward the health care professionals and increasing consumerism in health services (Roberts, 2002 p285) means that the general public are now more aware of the possibility of obtaining compensation for medical injuries and may be more likely to pursue this option (Harpwood, 2001). However, the impact of 'no win no fee' funding has still to be fully evaluated, although the citizens advice bureau has already discovered that the "actual number of claims for injuries following accidents has reduced since the conditional fee agreement was rolled out" (Sandbach, 2004).

Conclusions

Despite a thorough literature search using the Medline, Ebsco and Aditus databases, this chapter demonstrates that there appears to be very little published evidence that complaints management has a significant impact on litigation. Indeed, it is difficult to identify any single factor responsible for the increase in claims since 1996 (Harpwood, 2001) and subsequent fall in claims since 2001 (NHSLA, 2004a). It is reasonable to conclude that there are several factors which influence a claimant's decision to take a complaint or claim of negligence to court.

The development of risk management systems which address complaints management, incident reporting, education and training of staff, communication and information provision in conjunction with the civil justice reforms may all have an impact upon the incidence of civil actions in law.

It would be useful to understand in more detail exactly how these areas impact upon litigation and also to evaluate the effectiveness of the complaints process – not only from a claims perspective, but from a satisfaction viewpoint for both patients and staff members. There is little doubt that dealing with complaints currently causes stress and feelings of inadequacy amongst healthcare professionals whilst the perceptions of patients and their carers are less clear. Undoubtedly, this indicates a need for further research and service evaluation.

With increased knowledge, robust risk management systems and developments in the civil legal system, it is to be hoped that, in the longer term, fewer claims will be brought against health care professionals and NHS Trusts, and of those claims that are initiated, more will be settled out of court as a result of improved procedural changes in the conduct of litigation (Harpwood, 2001).

References

Adams A (2003.) *Law for Business Students*. 3rd ed. Pearson Education Ltd, Harlow.

Allsop J and Mulcahy L (2001). Dealing with clinical complaints. In Vincent, C. (2001). *Clinical Risk Management - Enhancing Patient Safety*. BMJ Publishing Group, London.

Ambady N, Laplante D, Nguyen T, Rosenthal R, Chaumeton N and Levinson W (2002). Surgeons' tone of voice: a clue to malpractice history. *Surgery* 132 (1): 5-9.

Amoore J and Ingram P (2002). Learning from adverse incidents involving medical devices. *British Medical Journal*. 3 August, 325: 272-275.

Barach P and Small S (2000). Reporting and preventing medical mishaps: lessons from non-medical near miss reporting systems. *British Medical Journal*. 18 March, 320: 759-763.

Cabinet Office Complaints Task Force (1995). *Putting Things Right*. HMSO, London.

Chapman E (2001). Claims Management. In Vincent, C (2001). *Clinical Risk Management - Enhancing Patient Safety*. BMJ Publishing Group, London.

Clinical Disputes Forum. (2001). *Guide to Mediation*. www.clinical-disputes-forum.org.uk

Clinical Disputes Forum (2002). *The Clinical Disputes Forum and its work*. www.clinical-disputes-forum.org.uk

Commission for Health Improvement (2002). CHI's combined annual report and accounts 2000-2001. www.chi.nhs.uk

Cooke J (1997). *Law of Tort*. (3rd edition). Pitman Publishing, London.

Cunningham W (2004). The immediate and long term impact on New Zealand doctors who receive patient complaints. *The New Zealand Medical Journal* 117 (1198), U972, 1175-8716.

Debell B (1997). *Conciliation and Mediation in the NHS – A practical guide*. Radcliffe Medical Press, Oxon.

Department of Health (1994). *Being Heard: the report of a review committee on NHS complaints procedures.* HMSO, London.

Department of Health (1996). *Guidance on implementation of the NHS complaints Procedure.* HMSO, London.

Department of Health (2000a). *The NHS Plan: a plan for investment, a plan for reform.* HMSO, London.

Department of Health (2000b). *An Organisation with a Memory.* HMSO, London.

Department of Health (2003). *Making Amends, a report by the Chief Medical Officer.* Department of Health Publications, London.

Department of Health (2004). *Guidance to support implementation of the National Health Service (complaints) regulations 2004.* HMSO, London.

Dyer C (2004). No win, no fee deals 'are failing claimants'. Guardian, 13th December 2004. www.guardian.co.uk/uk_news/story/0,3604,1372298,00.html

Eaton L (2003) Patient complaints procedure uncertain as health councils abolished. *British Medical Journal.* 8 February, 326:302.

Elliott C and Quinn F (2002.) *AS Law.* Pearson Education Ltd, Essex.

Golten D (2000) …..and justice for all?: clinical negligence claims. *British Journal of Healthcare Management* 6 (10): 479-483.

General Medical Council (2001). *Good Medical Practice* (3rd edition). www.gmc-uk.org

Gorney M (1999) The role of communication in the physician's office. *Clinics in Plastic Surgery* 26 (1): 133–141.

Harpwood V (2001). Clinical governance, litigation and human rights. *Journal of Management in Medicine* 15 (3): 227–241.

Health Service Ombudsman (2004). *Investigations completed October 2003 – March 2004.* www.ombudsman.org.uk

Hickson G, Federspiel C, Pichert J, Miller C, Gauld J and Bost P (2002). Patient complaints and malpractice. *Journal of the American Medical Association* 287 (22): 2951–2957.

Hinckley C (2003). Make no mistake: errors can be controlled. *Quality and Safety in Health Care* 12 (5): 359–365.

Jain A and Ogden J (1999). General Practitioners' experience of patients' complaints: qualitative study. *British Medical Journal* 318 (7198): 1596-1599.

Krause H, Bremerich A and Rustemeyer J (2001). Reasons for patients' discontent and litigation. *Journal of Cranio-maxillo-facial Surgery* 29 (3): 181-183.

Levinson W, Roter D, Mullooly J, Dull V and Frankel R (1997). Physician – patient communication. The relationship with malpractice claims among primary care physicians and surgeons. *Journal of the American Medical Association* 277 (7): 553–559.

Lord Chancellor (1999). *Access to Justice Act*. HMSO, London.

Longman (1984). *Dictionary of the English Language*. Longman Group UK Limited, England.

Mayberry M (2002.) The NHS complaints system. *Postgraduate Medical Journal* 78: 651–653.

Moore P, Adler N and Robertson P (2000). Medical malpractice: the effect of doctor-patient relations on medical patient perceptions and malpractice intentions. *The Western Journal of Medicine* 173 (4): 244-250.

Mulcahy L (1999). Being seen to be heard; reflections of a researcher on practice level of handling complaints. *Clinical Risk* 5 (3): 77-82.

Mulvaney L. (2004) Taking a closer look at the new complaints procedure. Healthcare Legal Update 42. www.weightmans.com

NHS Litigation Authority (1995). *Clinical Negligence Scheme for Trusts, reporting guidelines.* www.nhsla.com

NHS Litigation Authority (2004a). *About the NHS Litigation Authority.* www.nhsla.com

NHS Litigation Authority (2004b). *Factsheet 2: Financial Information.* www.nhsla.com/claims

Osbourne C (2002). *Civil Litigation*. Oxford University Press, Oxford.

Roberts G (2002). *Risk Management in Healthcare* (2nd edition). Witherbys, London.

Sandbach J (2004). No win, no fee, no chance. CAB evidence on the challenges facing access to injury compensation. www.citizensadvice.org.uk

Shaw R. (2004). Patient safety: the need for an open and fair culture. *Clinical Medicine* 4 (2): 128-131.

Shipman Inquiry (2005). *Shipman: The final report*. HMSO, Norwich.

Wilson, J (1998). Proactive risk management: effective communication. *British Journal of Nursing* 7 (15): 918-919.

Woolf, Lord Justice H (1996). *Access to Justice – final report*. www.dca.gov.uk

This page is intentionally blank.

7

The relationship between hospital acquired infection rates and the contracting out of cleaning services in the NHS in England - I

KIM HUDSON

"It may seem a strange principle to enunciate as the very first requirement in a Hospital that it should do the sick no harm."
Florence Nightingale (1859) *Notes on Hospitals*

Introduction

Healthcare associated infections (HCAI) are infections that are acquired as a direct result of healthcare. Hospital acquired infections (HAI) are those acquired during a stay in hospital.

The purpose of this chapter is to determine the extent of the association between the trends of rising HAI rates and the contracting out of cleaning services in England. I believe that there is indeed a fundamental relationship between the two, and will endeavour to discover whether the rise in infection rates is principally attributable to contract cleaning, or whether contract cleaning is one of a number of contributory factors.

The public are inundated with press and news reports about HCAIs, and there is a strong focus on the issue, with the main media attention being on the Methicillin Resistant Staphylococcus Aureus (MRSA) organism. MRSA has received most media coverage because of its resistance to antibiotics and its spread.

On 7 March 2005 the BBC reported that the number of antibiotic-resistant MRSA infections in England had fallen to the lowest since recording began, according to 'official figures'. The Labour government hailed these latest figures as a "turning point" in its efforts to combat the potentially deadly superbug. The opposition Conservative party, however, claimed that the figures failed to tell the full story and accused the government of "pre-election trickery." On 22 March 2005 the

BBC reported the death of a 36 hour old baby from MRSA, with the family accusing the hospital of a "cover up."

Media reports such as these often claim that there is a decline in cleaning standards, and government ministers have conceded this. In 2001, the then Health Secretary, Alan Milburn, wrote: "Standards of cleanliness have been poor in too many hospitals..." (NHS Estates, 2001)

On 23 March 2005 the BBC reported that the first outbreak of "winter vomiting bug", a stomach upset that causes sickness, mostly vomiting with some diarrhoea, had affected a hospital with 22 cases being reported. Winter vomiting bug is caused by a virus, technically named the Norwalk-like-virus or Small Round Structured Virus (SRSV). It is important to note that there are many other organisms causing HAIs that must also be considered (Fig 7.1).

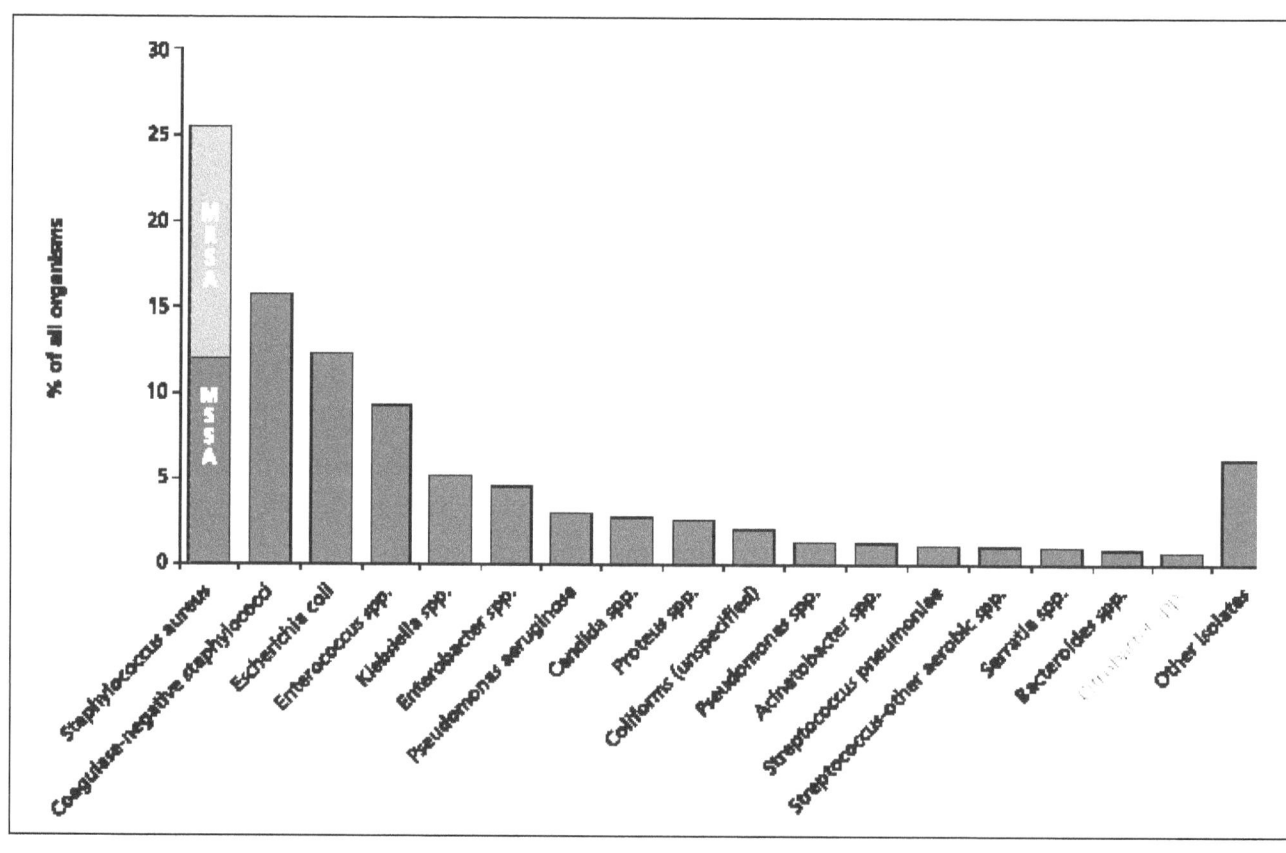

Fig 7.1 - Micro-organisms causing hospital-acquired bacteraemia
Source: Surveillance of Hospital-Acquired Bacteraemia in English Hospitals 1997 - 2002

HAIs are infections that are neither present nor incubating when a patient enters hospital. The majority – about three-quarters - of HAIs are caused by bacteria. Less common causes include specific infectious diseases such as viral gastroenteritis, influenza and, more rarely, tuberculosis and Legionnaire disease.

About nine per cent of inpatients have a hospital acquired infection at any one time, equating to around 100,000 infections a year (National Audit Office, 2000). A 1996 National prevalence study (Fig 7.2) found that urinary tract infections are the most common type of HCAI (23.2%) and bloodstream infections (septicaemia) have the highest mortality.

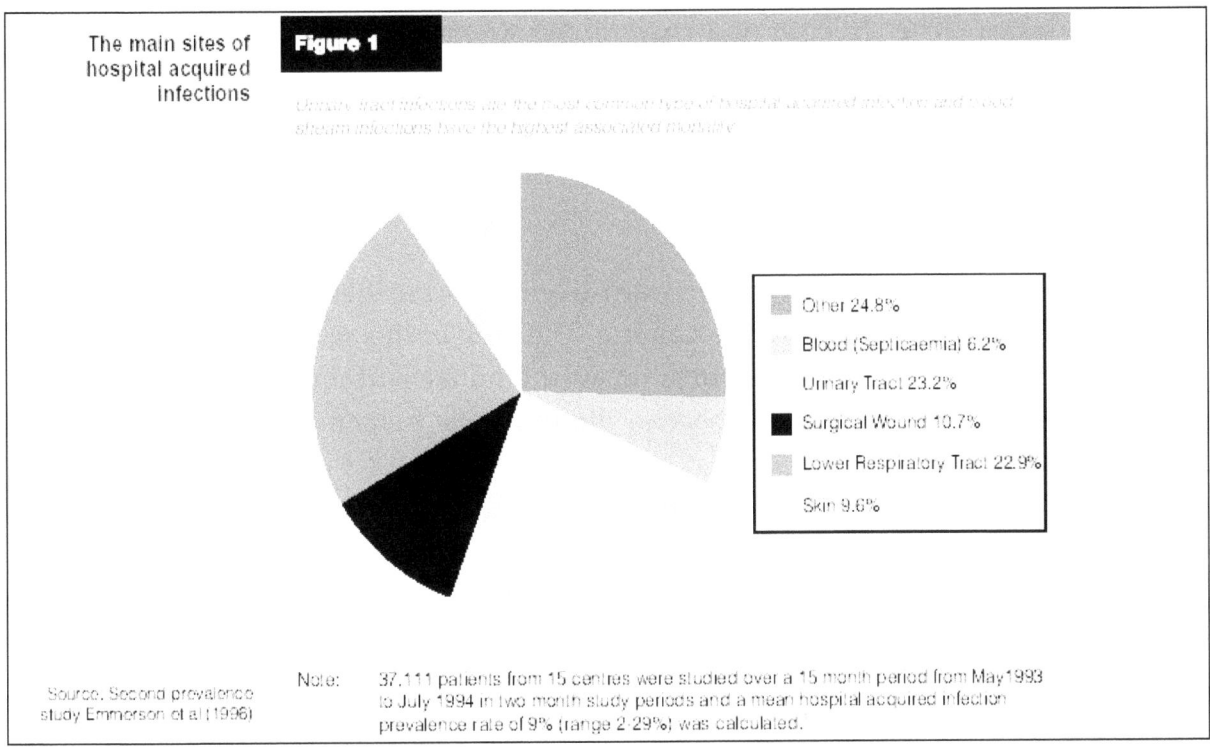

Fig 7.2: The main sites of hospital acquired infections. *Source*: National Audit Office, 2000.

A brief history of hospital acquired infection

On 4 November 1854, Florence Nightingale arrived at the Barrack Hospital in Scutari with a party of 38 nurses, just as 600 wounded soldiers were brought in from the battle of Inkerman in the Crimea. Her nurses firstly cleaned the whole hospital so there were no more germs and this helped to stop contamination and spread of disease. During her first winter at the Scutari Hospital, there was a 42% mortality rate among patients; however, by the end of the conflict, it had dropped to two per cent (Royal Statistical Society, 2003; Elliot, 2004).

Nightingale spent only two years nursing the sick and wounded, since a peace treaty was signed in 1856, and she returned home to England. Her subsequent writings on hospital planning and organisation had a profound influence in England and across the world. She believed that infection arose spontaneously in dirty and poorly ventilated places and, despite this mistaken belief, her far-sighted reforms continue to influence building design and the nature of modern healthcare (www.florence-nightingale.co.uk/).

Almost a century later in 1938, Balliere, Tindall and Cox , an English medical and scientific publishing house started to produce a series of textbooks for nurses. One of these texts, 'Aids to Hygiene for Nurses' (Funnell and Muir, 1950), states: "The strength of a chain is the strength of its weakest link – and ignorance of the laws of hygiene may cause, and often does cause, disease and injury to enormous numbers of innocent people." The publication further states that "As hygiene deals mainly with mundane matters, it is not always easy to inspire interest in it."

In England in the 1950s, domestic ward work was considered an "usual part of the daily routine of the nurse in training. The bulk of this work is done by ward maids and scrubbers, but the nurse must be familiar with methods of cleaning, so that she may efficiently supervise the domestic helpers..." (Houghton and Rosenheim, 1949).

In the 1980s, the public services were widely regarded as being fundamentally inefficient. 'Market testing' (also known as competitive tendering) was seen as the 'cure' for this inefficiency (Higgins and Roper, 2002). There was widespread privatisation of public services during the Conservative administration (1979 – 1997) and Compulsory Competitive Tendering (CCT) was introduced in 1980 with the ostensible aim of creating 'contestable markets' by opening up to competition activities that were previously immune from such competition. According to the Conservative government, this would reduce "waste, bureaucracy and over government" (Conservative Party, 1979 - cited in Higgins and Roper, 2002).

In 1983, the Conservative party, in their election manifesto, undertook to reduce the cost of administering the health service and, thereby, to release more funds to "improve patient services." This was to be achieved by asking health authorities to make the "maximum possible savings" by putting some services, including cleaning, out to competitive tender (Conservative Party, 1983; cited in UNISON, 2005). This increased pressure to save money increased the likelihood that the cheapest bidder would win the contract, and that insufficient weight would be placed on achieving quality outcomes (NHS Estates, 2004a). It also increased the likelihood that in-house services would be cut, since cleaning is labour intensive with around 90% of the cost of the contract being staffing costs. UNISON, Britain's biggest trade union, observed the inherent failure of the then government to recognise that cleaning was, and still is, a core "patient service" (UNISON, 2001).

In 1997, the new Labour government continued to share the previous Government's ambition of correcting the apparent failings of public sector organisation through applying private sector principles of efficiency and competition (Grimshaw et al, 2002). However, they did abolish the "compulsory" element of CCT, replacing it with 'best value' that required public bodies to consider factors other than cost alone. Contracting out of cleaning services in hospitals has, however, continued; and in 2001 UNISON estimated it accounted for 30% of cleaning services.

In 1996, a surveillance system was created, 'NINSS' (Nosocomial Infection Surveillance Service), with 102 voluntary participating hospitals collecting data, including demographic, number and rates for acute specialities, sources of infection, micro-organisms and facts about antibiotic resistance. This information can be used as a benchmark by hospitals nationally to measure their own performance, but does not relate any of the results with standards of cleanliness across the NHS, as there was previously no means of measuring this.

In 2000, independent Patient Environment Action Teams (PEATs) were established to review standards in acute hospitals. PEATs conduct annual reviews using specific criteria and defined protocols employing a five point-scale scoring mechanism which allows, for the first time, there to be an accurate picture of the standards of cleanliness across the NHS.

The NHS Plan (Department of Health, 2000) stated that there would be greater investment, national standards for cleanliness and a shift in contractual arrangements to ensure nurses can take the lead in ensuring wards are properly cleaned, as well as reintroducing the 'Matron'. Figure 7.3 is an extract from the NHS Plan outlining the new investment in NHS facilities to tackle infection risks.

> **Clean hospitals**
>
> **4.14** The new investment we are making will allow the NHS to get the basics right. Patients perceive a major deterioration in the cleanliness of hospitals since the introduction of Compulsory Competitive Tendering and the internal market. Patients expect wards to be clean, furnishings to be tidy. The new resources will allow for a renewed emphasis on clean hospitals.
>
> **4.15** As a result of this NHS Plan there will be:
>
> - over £30 million allocated immediately direct to hospital trusts to improve hospital cleaning. In future years, cash for cleaning will be distributed as part of the normal allocation process
> - a nation-wide clean-up campaign throughout the NHS starting immediately. All patient areas, visitor toilets, outpatients and accident and emergency units will be thoroughly cleaned and kept clean. Chairs, linen, pillows, furniture, floor coverings and blinds which are dirty will be cleaned. Those beyond repair will be replaced
> - every hospital will have an unannounced inspection of its cleanliness, by a specialist inspection team including patients within the next 6 months. The results will be made available to the local media
> - national standards for cleanliness will form part of the NHS Performance Assessment Framework. Every hospital's performance will be measured against these standards by the end of 2000
> - ward sisters and charge nurses will have the authority to ensure the wards they lead are properly cleaned. Hospital domestics will be fully part of the ward team – and respected for the important work they do
> - NHS trusts will have to adjust contracts with external cleaning companies to ensure nurses can take the lead in ensuring wards are properly cleaned where necessary
> - every NHS trust will nominate a Board member to take personal responsibility for monitoring hospital cleanliness, and will report to the Board following regular check ups. The Patients' Forum in each trust will monitor standards and use unannounced inspections to do this
> - the independent NHS watchdog, the Commission for Health Improvement, will inspect and report on this aspect of NHS care.

Figure 7.3 - Extract from The NHS Plan (2000), Chapter 4: Investing in NHS facilities

The Department of Health's 2003 report 'Winning Ways' suggested seven 'action areas' in recognition "of concerns that standards of cleanliness in some hospitals were not as good as patients and the public had the right to expect."

'A Matron's Charter: An Action Pan for Cleaner Hospitals' (Department of Health, 2004) sets out 10 broad commitments that should be adopted, emphasising ownership and teamwork. This report acknowledged that "cleanliness is everyone's responsibility" and recognised that nurses and midwives provide an "unbreakable strand of continuity, delivering a 24-hour service", so they should bear the responsibility for making sure hospitals are clean, and challenge poor practice and recognise achievement across all disciplines.

Today, nearly 150 years after Florence Nightingale returned to England, hospital cleanliness continues to be a much debated issue, with various initiatives over the last decade aimed at solving the problem of rising HAI's demonstrating variable degrees of success.

Available literature

There are presently newspaper stories in the National tabloids, broadsheets and on-line appearing almost daily linking hospital acquired infection and hospital cleanliness. There can be political bias concealed within, together with hysteria-inducing headlines using political and emotive wording; "Downing Street denies NHS funding rift…"; "Hospitals clean up their act."; "Lack of resources hampers superbug fight, nurses say"; "'Worst' hospital disputes MRSA infection figures."; "Hospitals need a good scrub" etc. (e.g. see http://society.guardian.co.uk).

There is a plethora of acclaimed UK-based electronic websites containing information on hospital acquired infections *and* cleaning, both governmental and non-governmental. The governmental websites display Department of Health (Department of Health) publications and circulars, such as 'Getting ahead of the Curve' (Department of Health, 2002) 'Winning Ways' (Department of Health, 2003) and 'Towards Cleaner hospitals and lower rates of infection' (Department of Health, 2004). All of these reports tend to focus on the increased antibiotic resistance as a key factor in the rise of hospital acquired infections, and they list the key Department of Health initiatives which have been launched over the years to attempt to reduce Hospital Acquired Infection, by providing guidance on antibiotic prescribing, improving cleaning standards, and improving infection surveillance. But none makes a direct correlation with contracting – simply acknowledging that infection rates are rising and cleaning standards falling.

There are various other informative documents, from non governmental professional bodies' briefing papers and audit reports, and from both clinical and non-clinical recognised bodies such as the Royal Colleges and Unions, frequently reflecting on the association between infection and cleaning, and sharing research and findings and making recommendations on how to deliver improved standards of cleanliness (UNISON, 2005a). There were seemingly only two reports, both from the Union which directly compare infection control in hospitals and the contracting out of public services (UNISON, 2005; UNISON 2005b). The author considers there is an element of political bias which pervades many of the UNISON reports, making it challenging to extract impartial information. The National Audit Office (NAO), who are independent of the government and scrutinise public spending on behalf of Parliament have published two reports (NAO, 2000; NAO, 2004), which seem to have greater detail and more specific recommendations relating to all aspects of hospital acquired infection. Indeed the NAO report on the 'Management and Control of Hospital Acquired Infection in Acute NHS Trusts in England' lists thirty four recommendations, which are subsequently described in finer detail.

There are numerous Medical and Nursing Journal articles (Mayor, 2000; Curran, 2001; Beckford-Ball & Hainsworth, 2004; Castledine, 2004; Jenkins, 2004; Rayner, D. 2004;) with many focusing on personal hygiene and handwashing (Duffin & Scott, 2000; Parker, 2004; Teare et al, 2004;) whilst another argued that the infections were incorrectly categorised as 'hospital-acquired' where they ought to have been classified as community acquired (Leesens et al, 2005).

Other articles examine the success of the surveillance systems (Coello et al, 2001; Kelsey, 2001; National Audit Office, 2003; Waters, 2005).

Newspaper articles can appear biased and lack credibility if there is no reference to research in support of its piece. Appraising the quality of the research referenced within the piece is a difficult undertaking and requires experience of critical appraisal. The National electronic Library for Health (www.nelh.nhs.uk) recognised this lack of reliability and commissioned the University of York's Centre for Reviews and Dissemination (CRD) to assess the reliability of both the journalists' reporting of health stories and the research on which they are based. Their "Hitting the Headlines" project allows clinicians and the public alike to be better informed about the accuracy of reporting in the press, with the aim of making summaries live within 48 hours of newspaper publication.

Mulholland (2005) published an article on the Guardian Unlimited website on March 8th 2005 detailing the findings of a Nursing Times survey of 2000 nurses who were bemoaning a lack of cleaning facilities including uniform laundering. There are approximately 660,000 registered nurses working in the UK (Nursing and Midwifery Council, 2004), so the survey represented approximately 0.3% of UK nurses' opinions. This is one of numerous newspaper on-line stories which can cause anxiety amongst the public, and it could be debated whether, if the NeLH had appraised this story, they would have denounced the findings as insignificant due to the small number of nurses surveyed. The article did go into further detail on some UNISON claims that the numbers of cleaners had been halved following the outsourcing that has taken place. This is questionable, since in 2003 a UNISON press statement remarked about the cleaners' numbers being halved under Conservative government administration, stating that there were 88,307 cleaners working in the NHS, however the Department of Health claim there are no actual statistics on the numbers of cleaners working in the NHS as statistics are no longer collected under specific occupational groups.

Regardless of whether there has been a decline in the number of cleaners over the last two decades, which may or may not have had a significant impact upon the rise in rates of infections, in a NAO Report (2000) it was noted that trust membership (and indeed attendance) of Infection Control Committees was variable. It could be construed that NHS trusts previously regarded the cleaner's role as perfunctory as they did not appear to consider involving a cleaning supervisor/manager representative with their infection control programme (see Fig 7.4), although this may be because Department guidance issued in 1995 omitted to include them in their recommended list of membership of the Infection Control Committee.

Figure 8 Membership of NHS Trust's Hospital Infection Control Committees and frequency of attendance

This shows that the composition of Hospital Infection Control Committees vary as do the patterns of attendance

Departmental Guidance Suggested membership	Number of NHS Trusts who specified membership	Always/ Sometimes attends[1]	Rarely /Never attends	Didn't answer
Chief Executive[2]	22 (11%)	14 (64%)	5 (23%)	3 (14%)
Chief Executive Representative[2]	138 (65%)	123 (90%)	6 (4%)	9 (7%)
CCDC	174 (81%)	156 (89%)	9 (6%)	9 (6%)
Infection Control Doctor	205 (96%)	190 (93%)	-	15 (7%)
Infection Control Nurse	209 (98%))	194 (93%)	-	15 (7%)
Occupational Health Physician[2]	126 (59)	98 (78%)	19 (15%)	9 (8%)
Occupational Health Nurse[2]	163 (76%)	151 (93%)	3 (2%)	9 (6%)
Infectious Disease Physician	38 (18%)	30 (79%)	6 (16%)	2 (6%)
Senior Clinical Medical Staff	182 (85%)	142 (78%)	25 (14%)	15 (8%)
Nurse Executive Director	132 (62%)	110 (84%)	8 (6%)	14 (11%)
Representative from other hospitals covered by HICC (n=120 NHS Trust)	62 (29%)	26 (42%)	6 (10%)	9 (15%)

Notes: 1. Not all respondents who answered membership question detailed frequency of attendance

2. Either / or both can be members

Source: National Audit Office census, analysis of the membership of the 215 trusts who stated that they had a Hospital Infection Control Committee

Figure 7.4 - Source: National Audit Office census, analysis of the membership of the 215 trusts who stated that they had a Hospital Infection Control Committee. National Audit Office; The Management and Control of Hospital Acquired Infection in Acute NHS Trusts in England

A later report by UNISON, 'Cleaners' Voices' (2005), endeavoured to make sure that "the voice of the experts – the cleaners themselves – is heard by the policy-makers," and they went on to suggest 10 key steps, with two key steps focusing on 'effective teams' and 'respect and improving communication' stating that

"Cleaner Hospital" teams should include representatives of cleaning staff, and they must be viewed and treated as part of the healthcare team.

Davies' independent 2005 report for UNISON (UNISON, 2005a) also described how contracting out separates the cleaners from the rest of the ward team and undermines the team-based approach, altering the cleaners' attitudes to their jobs, and damaging their general commitment to the goals of the organisation. The author suggests the team-based approach has as much to do with the 'leadership' on the wards as it does whether the cleaners are contracted or in-house. I don't understand the preceding sentence. Indeed, the Matron's Charter, developed in October 2004 following the recommendation within the Department of Health's policy "Towards cleaner hospitals and lower rates of infection" reinforces the role of Matron as a role model who must lead by example and ensure that all staff are empowered to embrace good practice, whether they be in-house or contracted.

Another initiative to attempt to improve the working conditions and enhance the role of the cleaners was to introduce a 'Ward Housekeeper' service. This was part of the original remit of the Patient Environment Teams, which were established in 2001. These ward housekeepers would be "part of the ward team", and provide for the non-clinical needs of the patients, therefore leaving the nursing staff free to do the job they were trained for – nursing the patients back to health - and be responsible to the ward manager (NHS Estates, 2004b)

UNISON's 'independent' report (UNISON, 2005a) also pointed out that "it hardly seemed a coincidence" in one of the government's first surveys of hospital cleanliness, in the 2001 PEAT Review, that 20 out of 23 hospitals identified as having the worst standards of cleanliness in the NHS employed private contractors. Despite this being a 2005 report, it failed to mention the improvement in the 2004 PEAT review (Fig 7.5), which demonstrated substantial improvements (but a deterioration, when compared with 2003) with an updated system to allow more scope for differentiation between hospitals, and the number of 'elements' (areas assessed) being increased from 18 to 24 to ensure the results are based on a greater range of services, providing an even more accurate assessment of a hospital's performance. The number of hospitals being reviewed has also increased by nearly 50% since initial reviews in 2000.

	Red (poor)	Yellow (acceptable)	Green (good)	Excellent	No of Hospitals
Autumn 2000	253 (35.5%)	297 (41.7%)	163 (22.3%)		713
Spring 2001	42 (6.1%)	368 (53.4%)	279 (40.5%)		689
Autumn 2001	0 (0%)	387 (56.3%)	300 (43.7%)		687
2002	0 (0%)	317 (40%)	464 (60%)		781
2003	0 (0%)	186 (21.3%)	686 (78.7%)		872
2004	24 (2%)	583 (49%)	456 (38.5%)	118 (10%)	1181

Figure 7.5 - Adapted from Clean Hospitals, PEAT Results

The Nosocomial Infection National Surveillance Service (NINSS), created in 1996 and sponsored by the Department of Health, started to report on Hospital Acquired Infection in 1997, observing that cumulating data over time would enable more precise estimates of the incidence of infection to be calculated (Public Health Laboratory Service, 2002). However since the system was designed to provide only *participating* Hospitals with comparative data, thus excluding those who did not participate, any results or trend analysis would be limited. The NAO concurred with this, and the 'Management and Control of Hospital Acquired Infection in Acute NHS Trusts in England' report in 2000 stated that surveillance needed to be done more effectively, commenting that there were limited comparable data and wide variations in the extent of dissemination of surveillance results.

In 2002, the House of Lords Select Committee on Science and Technology established a subcommittee to carry out an inquiry into diagnosis, treatment, prevention and control of infectious diseases, issuing a call for evidence before publishing their final report in July 2003 (House of Lords, 2003). They noted that the completeness of surveillance is ad hoc, as it does not depend on a legal requirement, however, even those aspects that do have a legal requirement (for example symptoms of food poisoning) are not adequately reported.

The NAO responded to the call for evidence and suggested that surveillance should be made mandatory (National Audit Office, 2003). Their report contained 11 clearly defined recommendations concerning surveillance, stating that research shows that "surveillance, involving data collection, analysis and feedback of results to clinicians, is central to detecting infections, dealing with them, and ultimately reducing infection rates."

Despite this detailed submission, the House of Lords report's conclusions were non-specific, stating that concerns regarding surveillance could be addressed by designing better and more innovative information systems and improving funding to laboratories. Two years later, in 2004, another report by the NAO, ordered by the House of Commons, observed that despite earlier reports and recommendations, change continues to be constrained by the lack of data and limited progress in implementing a national mandatory surveillance programme.

UNISON's independent report (UNISON, 2005a) lists an abundance of specific problems and difficulties associated with contracting out, including much discussion about the relationship between the contracted employers and their fellow public service workers. They state that there can be inflexibility, and that monitoring their work can diminish trust, as well as there being increased sickness, absence, and recruitment and retention issues. The author accepts these issues can be counter productive and demotivating for staff; and, although there is evidence within the report to suggest that staff turnover amongst external contractors is higher, the problems and difficulties remain similar for both in-house *and* contracted staff.

Today, all hospitals are monitored equally, and if an inspection encounters a cleanliness concern, recommendations are made, detailed action plans are drawn up and progress is measured. ALL staff would be expected to abide by these recommendations, whether contracted or in-house. Most contracts with private sector providers have penalty clauses, although the 2005 independent UNISON report acknowledges that there are strong pressures not to impose these for a variety of reasons, mainly as it would "almost certainly damage, perhaps irrevocably, the relationship between purchaser and provider."

There have been numerous governmental initiatives over the last five years that have attempted to redress the balance and better manage prevention of infection. The 'Getting ahead of the curve' report (Department of Health, 2002) primarily scoped the threat and focussed on a strategy to combat a wide range of global infectious diseases including methicillin resistant staphylococcus aureus and tuberculosis, amongst others. They proposed 12 actions, including proposing to merge existing bodies and create a new national agency to act as a source of national expertise and strengthening the surveillance system.

The 'Winning Ways' report (Department of Health, 2003) followed this, reiterating the actions of the 2002 report, but amalgamating the actions into seven 'Action Areas', emphasising that local infrastructure and systems plus senior management commitment is vital. So they proposed that Strategic Health Authorities should be accountable for ensuring that NHS performance management arrangements are aligned to achieve the objectives within the report; and part of the local audit arrangements should include assessment of adherence to standard (universal) precautions to reduce the transmission of HCAIs. They would be asking the Healthcare Commission to give priority to assessing NHS performance in reducing healthcare associated infection.

Discussion

It is vital to continuously appraise the reliability of newspaper articles, but especially so when warring political parties are seizing on an issue to score political points when an election is looming. Thus, whilst it is beneficial for the Press to report on hospital acquired infection, and keep it high on 'the public agenda', it is disappointing to concede that the information within may not always be accurate, unless there is direct reference to research. Thus the public may be misinformed and the impact may be more evident in the ballot box than in our hospitals.

Data on hospital acquired infection was not systematically collected until the 'NINSS' surveillance system commenced in 1996. Thus, there are less than 10 years of comparable data, with much of those being inconsistently collected. The 'NINSS' data collection is currently unconnected with the independent programme of hospital 'PEAT' reviews. If they were co-ordinated they might give us more robust correlated information providing evidence of the number of infection rates proportionate to the cleanliness of the hospital being reviewed.

Historically, there was a tendency to focus on treatment rather than prevention, with infection control being largely reactive in nature (NAO, 2005). The various initiatives addressing cleanliness and HCAIs appear to have merely reiterated the same or similar actions and, not until the 2003 Department of Health report 'Winning Ways' stipulated that the Healthcare Commission reviews would include assessments of trusts' performance related to reducing healthcare associated infections, did the importance of addressing cleanliness and infection control assume a new intensity.

Trusts must then have welcomed the Department of Health 2004 report 'Towards cleaner hospitals and lower rates of infection' which endeavoured to empower patients, nurses and "all NHS staff" to "bring everywhere up to the level of the best by sharing good practice." However, the report again merely reiterated many of the initiatives and reports of previous years and the only new message was that "we need to make faster progress."

Whilst acknowledging the importance of sharing best practice using evidence to support the issue of cleanliness and infection, the author speculates whether there would be less antibiotic resistant infectious disease if doctors had heeded advice when the first guidelines for dealing with MRSA were issued in 1986, as they "doled the exciting new medicines out in large amounts, grateful for a weapon against previously serious conditions, but also in response to huge patient demand" (Bowlby, 2005). Kakkilaya (2005) suggests four reasons why doctors still continue to over-prescribe antibiotics:

- Lack of confidence – courage to avoid unnecessary prescriptions
- Peer pressure – their colleague may prescribe and get 'credit for the cure'
- Patient pressure – patients who insist to 'get better at the earliest'
- Company pressure – pharmaceutical companies plying their wares

Since it has been implied that contracted cleaners may provide a lower quality service than their in-house colleagues (UNISON, 2005a), where contracts are due for renewal they need to be made specific to ensure acceptable levels of cleanliness and compliance with targets. Where previously standards were not audited, and therefore sanctions for poor performance were not imposed, there is now limited defence for organisations with the detailed guidance which now exists (NHS Estates, 2004a) where it advises that tenders are assessed according to specified exclusion, selection and award criteria, designed to eliminate companies who fail to meet the selection criteria in relation to their capacity of providing a quality service.

Whilst there are many reports cited within this study attributing the rise in infection rates to contract cleaning, less blame seems to be apportioned to the increased throughput of patients through the hospital beds. Concern has been raised that the success in reducing waiting lists and a rise in bed occupancy rates have led to a higher rate of MRSA (Department of Health, 2004; Simmons et al, 2005).

The Kings Fund commissioned a review, "An Independent Audit of the NHS under Labour" in March 2005, which highlighted four "big issues", with one being the declining number of beds (Figure 7.6) with increasing occupancy rates rising from 84% in 2000/01 to 85.8% in 2003/04, higher than other developed countries. The bed occupancy is high as hospitals push to reduce waiting lists and achieve targets set by the Government. The Department of Health concede that the data demonstrating the rates of MRSA bacteraemias published by the Health Protection Agency Communicable Disease Surveillance Centre are "not straightforward" due to the way that Hospitals calculate their bed occupancy rates (Department of Health, 2005).

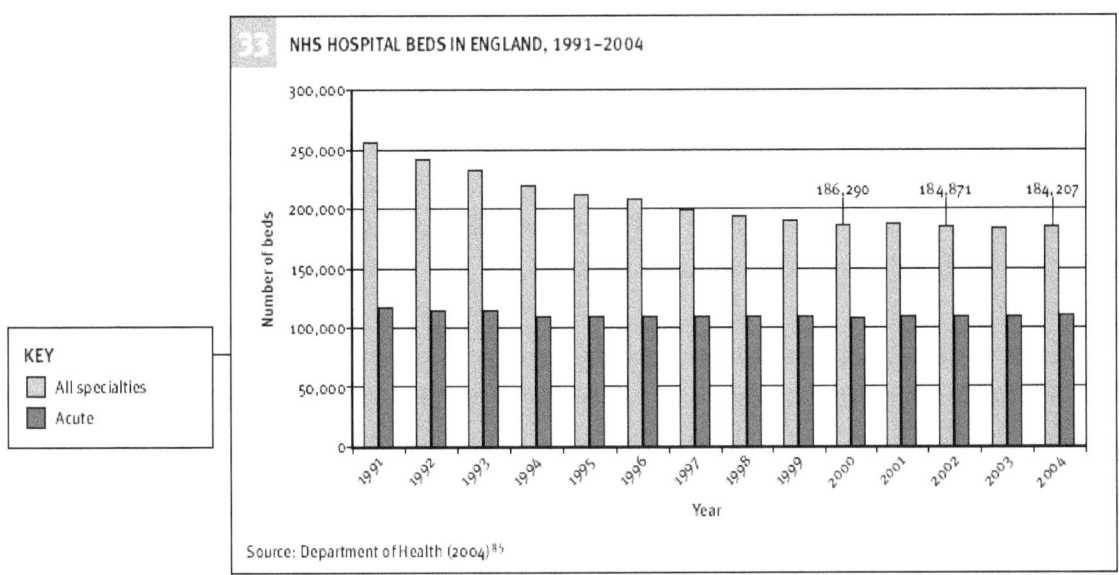

fig 7.6 - Source: An Independent Audit of the National Health Service under Labour (1997–2005); Kings Fund March 2005

Concluding comments

There remains relatively limited and contradictory evidence on the impact of selective contracting on efficiency and equity at the facility and/or at the health system level, which highlights the need for extensive additional research on the effects of the various reforms and initiatives now being implemented.

This review has also shown that many of the theoretical claims on the basis of which contracting reforms are argued to improve efficiency themselves remain ambiguous. UNISON commissioned an 'independent' report on 'Hospital contract cleaning and infection control' in January 2005 which stated in its conclusion that:

"Contracting out cleaning services is clearly not the only reason for the spread of HAI in general: poor hand washing practices and antibiotic policies, excessive movement of patients, shortage of beds and rapid patient throughput and higher than recommended levels of bed occupancy to meet performance targets...."

A month later a press release from the same Organisation, by Dave Prentis, the general secretary of UNISON stated:

"It's a bit rich to hear the Tories flinging accusations at this government. It was their policies that let superbugs loose in our hospitals through cutting and privatising cleaning services. Hospital contracts went to the cheapest bidder and we are all paying the price – dirt isn't cheap."

Whilst the Secretary of State for Health at the time, Alan Milburn, acknowledged that "compulsory competitive tendering has gone – because it failed to raise standards" (NHS Estates, 2001), he did not state that this was clearly the *only* reason standards had fallen.

Despite the proliferation of reports and "initiatives", the Department of Health acknowledges that progress is slow; and whilst it accepts that cleanliness and infection control are "closely linked" it states that there are "more important distinctions", since preventing infections requires more than "simple cleanliness."

Funnel and Muir (1950) observed that staff were less inspired to be concerned with the "mundane" matter of hygiene, and in 2000 the NAO observed that "infection control was not high enough on the agendas of NHS trust chief executives." Is this because it still considered "mundane"?

Surveillance data on Hospital Acquired Infection rates has been variable and diverse, and the NAO (2000) recommended developing a combination of formal mandatory targeted and selective surveillance with improved feedback to allow hospitals to take appropriate action to improve quality of care. The data would be more meaningful if other variables were accounted for, namely type of hospital – acute (with and/or without an Emergency Department), size of hospital, age of hospital, bed occupancy, type of patients treated, in-house infection control processes and cleaning arrangements. The World Health Organisation (2002) has also listed factors which influence the interpretation of antimicrobial resistance data, stating that the results depend on the microbial agent, patient susceptibility, environmental factors and bacterial resistance; thus there is still much work to be done to ensure that surveillance data is more meaningful, allowing parallels to be drawn between hospital cleanliness and infection rates.

Dr Norman Simmons, Emeritus Consultant Microbiologist at Guy's and St Thomas' Hospital Trust and several of his prominent specialist Microbiology colleagues (2005) wrote to the Times newspaper expressing their concern that many people believe that "cleaning hospitals and more handwashing" will solve the problem.

Perhaps, if less duplicative and ambiguous reports were produced, and more attention focused on *how* all the valid recommendations on hospital cleanliness and infection control could be implemented, then perhaps there would be some real achievements and results, as Florence Nightingale once suggested and was quoted in Cecil Woodham-Smith's 1951 biography: "You ask me why I do not write something.... I think one's feelings waste themselves in words, they ought all to be distilled into actions and into actions which bring results."

References

Beckford-Ball J and Hainsworth T (2004). The control and prevention of hospital-acquired infections. *Nursing Times* 2004, 20 Jul, 100(29), pp 28-29.

Bowlby C (2005). How misuse of antibiotics backfired. Monday 14th February 2005. http://news.bbc.co.uk/2/hi/health/4264121.stm

Castledine G (2004). Will infection be reduced in the coming year? *British Journal of Nursing* Jan 8-Jan 21, 2004. Vol. 13, Iss. 1; p. 59.

Coello R, Gastmeier P, de Boer (2001). Surveillance of hospital-acquired infection in England, Germany, and The Netherlands: will international comparison of rates be possible? *Infection Control & Hospital Epidemiology* Vol. 22, Iss. 6; pp 393 – 397.

Curran, E (2001). Reducing the risk of healthcare acquired infection. *Nursing Standard* Vol 16 No. 1, pp45-52.

Dancer S J (2004). How do we assess hospital cleaning? A proposal for microbiological standards for surface hygiene in hospitals? *Journal of Hospital Infection* Vol 56; Issue 1; January 2004; pp 10-15.

Department of Health (2002). *Getting ahead of the Curve - A strategy for combating infectious diseases (including other aspects of health protection).* A report by the Chief Medical Officer. www.dh.gov.uk

Department of Health (2003). *Winning Ways - Working together to reduce Healthcare Associated Infection in England.* www.dh.gov.uk

Department of Health (2004). *Towards cleaner hospitals and lower rates of infection - A summary of action.* www.dh.gov.uk

Department of Health (2005). *MRSA Surveillance System – results.* Published 07/03/05. www.dh.gov.uk

Duffin C and Scott G (2000). Ministers crack down on hospital-acquired infection. *Nursing Standard* Vol 15, No. 5 p7.

Elliot J (2004). The multi-faceted Lady with the Lamp. http://news.bbc.co.uk/2/hi/health/3943997.stm

Funnell E M and Muir C (1950). *Aids to Hygiene for Nurses*. 4th edition. Balliere, Tindall & Cox, London.

Higgins P and Roper I. Does Best Value Offer a Better Deal for Local Government Workers than Compulsory Competitive Tendering? Discussion Paper; Middlesex University HRM.
http://mubs.mdx.ac.uk/Research/Discussion_Papers/Human_Resource_Management/dpaphrmno8.pdf

House of Lords (2003). *Fighting Infection.* Select Committee on Science and Technology, Session 2002-03. 4th Report.

Jenkins L (2004). The prevention of Clostridium difficile associated diarrhoea in hospital. *Nursing Times* 29 June 2004, Vol 100, No. 26, pp 56-57.

Kakkilaya B S (2005). Why do doctors over-prescribe antibiotics?
http://rationalmedicine.org/drugs/antibiotics_abuse.htm

Kelsey T (2001). Which doctor? - need for standards to measure quality of British health care. *New Statesman* January 15 2001 www.newstatesman.com

Kings Fund (2005). An Independent Audit of the NHS under Labour (1997-2005). Commissioned by The Sunday Times. March 2005.

Lesens O, Hansmann Y, Brannigan E, Hopkins S (2005). Healthcare-associated Staphylococcus aureus bacteremia and the risk for methicillin resistance: is the Centers for Disease Control and Prevention definition for community-acquired bacteremia still appropriate? *Infection Control & Hospital Epidemiology* Vol. 26, Iss. 2; pp 204 -209.

Mayor S (2000). Hospital acquired infections kill 5000 patients a year in England. *British Medical Journal* (International edition). Vol 321, No. 7273 pp 1370.

Mulholland, H. (2005). Lack of resources hampers superbug fight, nurses say.
http://society.guardian.co.uk/nhsperformance/story/0,,1432973,00.html

National Audit Office (2000). *The Management and Control of Hospital Acquired Infection in acute NHS Trusts in England*. The Stationery Office, London.
www.nao.org.uk/publications/nao_reports/9900230.pdf

National Audit Office (2003). *Memorandum - House of Lords Science and Technology Select Committee inquiry "Fighting Infection"*, Session 2003-2004 www.publications.parliament.uk/pa/ld200203/ldselect/ldsctech/138/138w10.htm

National Audit Office (2004). *Improving patient care by reducing the risk of hospital acquired infection; A progress report*. Report by the Comptroller and Auditor General HC 876 Session 2003-2004. www.nao.org.uk/publications/nao_reports/03-04/0304876.pdf

NHS Estates (2001). *National Standards of Cleanliness for the NHS*. London: Department of Health.
www.hcsu.org.uk/index.php?option=com_docman&task=cat_view&gid=84&limit=5&limitstart=35

NHS Estates (2004a). *Revised Guidance on Contract for Cleaning*. December 2004. www.dh.gov.uk

NHS Estates (2004b). Improving the Patient Experience - Ward Housekeeping. www.dh.gov.uk

Nursing and Midwifery Council (2004). *Statistical analysis of the register 1 April 2003 – 31 March 2004*. www.nmc-uk.org

Parker L (2004). Infection control: maintaining the personal hygiene of patients and staff. *British Journal of Nursing* 13(8), pp 474-478.

Prentis D (2005). Press Release: UNISON says Think Cleaners on NHS Think Clean day.
www.unison.org.uk/healthcare/cleanerhospitals/pages_view.asp?did=1744

Public Health Laboratory Service (2002). Surveillance of Hospital-Acquired Bacteraemia in English Hospitals 1997 – 2002. A national surveillance and quality improvement programme in partnership with the Nosocomial Infection National Surveillance Service (NINSS).

Rayner D (2004). Tuberculosis and HIV infection: minimising transmission. *Nursing Standard* Vol 19, No. 4, pp 47-53.

Royal Statistical Society (2003). *A career as a statistician in the health service*. www.rss.org.uk

Simmons N A and others (2005). Hygiene not key to fighting MRSA. Times on-Line. www.timesonline.co.uk/article/0,,59-1509376,00.html

Teare L, Cookson B, Stone S, (2001). Hand Hygiene. *British Medical Journal* (International edition). London. Vol. 323, Iss. 7310; pp. 411–412.

UNISON (2001). Supplementary memorandum (PS 33A). Minutes of Evidence. The Role of the Private sector in the NHS. Select Committee on Health. First Report.

UNISON (2003). Press release. Hospital super-bug busting needs more cleaners. www.unison.org.uk/asppresspack/pressrelease_view.asp?id=388

UNISON (2005a). Hospital contract cleaning and infection control. An independent report from Steve Davies of Cardiff University, commissioned by UNISON.

UNISON (2005b). Cleaners' Voices. Interviews with hospital cleaning staff.

Waters A (2005). Operation clean-up. *Nursing Standard* Jan 26-Feb1, 2005; Vol 19; pp 18-19.

World Health Organisation (2005). *Prevention of hospital acquired infections: A practical guide* (2nd edition). www.who.int

This page is intentionally blank.

8

The relationship between hospital acquired infection rates and the contracting out of cleaning services in the NHS in England - II

JAYNE HARTLEY

Introduction

The intention of this chapter is to discuss whether the contracting out of cleaning services in the National Health Service (NHS) has had a significant impact upon the rise in the rates of hospital acquired infections.

Before commencing this discussion, the history of contract cleaning in the NHS will examined. This will include an understanding of its inception; the subsequent impact upon numbers of cleaners working in the NHS; and a review of the most recent guidance issued by NHS Estates (2004a), which provides, at the time of writing, the latest direction on contracting for cleaning. The incidence and costs of hospital acquired infections in the NHS is then explored. This will address key terminology and include reference to national data collated by the National Audit Office (NAO) and the Department of Health (DH). The assessment of hospital cleanliness will also be briefly discussed.

The factors which may impact upon the rates of hospital acquired infections, including the possible effects of the contracting out of cleaning services within the NHS and other issues, will then be considered. This will include topics which relate directly to infection control such as education and training, hand washing, audit and surveillance of hospital acquired infection, the strategic management of hospital acquired infection and funding arrangements for infection control. The chapter will also review other subjects which may not at first glance appear to impact on the incidence of hospital acquired infections, such as new roles and responsibilities for healthcare staff, antibiotic prescribing, hospital design and the management of medical devices and equipment.

The history of contract cleaning in the NHS

In 1983, the Conservative party published its general election manifesto in which it declared its intention to release more money for patient care by reducing the costs of administering the NHS. To meet this aim, health authorities were required to "to make the maximum possible savings by putting services like laundry, catering and hospital cleaning out to competitive tender" (Conservative Party 1983). Prior to 1983, cleaning services for NHS hospitals were provided by NHS cleaning staff, employed and managed directly by the hospital in which they worked. Only two per cent of expenditure on NHS cleaning went to contractors in 1982-1983 – and this was spent on office cleaning (Milne 1993).

In 1983, the Conservative Government also abolished the 'Fair Wages Resolution' (Unison 2005a). This resolution required firms undertaking contracts for the public sector to observe fair labour standards, and to follow the wage structure for the equivalent public sector worker (EMIRE, 2005). Abandoning this perceived 'model employer' approach (Unison 2005a) eased the way forwards for private companies to compete for cleaning contracts with in house teams by reviewing the wages and terms of employment for their employees. This action also reinforced the intention of Margaret Thatcher's Conservative party to confront the unions and fight the power they had at that time (Jones, 2004).

Information on the extent to which cleaning services have been contracted out is not held centrally by the DH (Hansard, 2000), although Unison (2003) estimates it to be 30% of cleaning services. The provision of contract cleaning is dominated by four main companies which are estimated to have 51% of all NHS contracted out cleaning services (Unison 2003). These companies have gone on to offer combinations of services to the NHS, including catering, portering and security services, thereby making themselves increasingly integral to the functioning of the NHS – and making it less easy for the NHS to revert to non-contractual arrangements.

There has been much debate about whether savings have been generated following the introduction of contract cleaning. For example, Domberger et al (1987), as cited by Unison (2005a) and Milne and McGee (1992), suggest that contract cleaning has led to savings of at least 20%, whilst a more recent study undertaken by Milne and Wright (2004) suggests that previous studies have over-estimated the cost-savings associated with competitive tendering. There can be little doubt that any savings made have come at the expense of cleaning staff – either by reducing wages or reducing numbers, since it is estimated that staff costs account for 93% of the costs of cleaning, with the remaining seven per cent going on cleaning equipment, materials and consumables (Unison, 2005a). It is claimed that the number of hospital cleaners has been nearly halved over the past 20 years (Unison, 2005b), falling from 100,000 in 1984 to 55,000 in 2003/2004 (Revill, 2005). The DH accept the accuracy of these figures, but claim that the fall in the number of NHS cleaners is partly due to a 20 percent reduction in the size of the NHS estate – meaning there is less to be cleaned (Revill, 2005).

In 2001, the Labour government ended the 'compulsory' element of compulsory competitive tendering following an acknowledgement by the Secretary of State that this arrangement had failed to raise standards (NHS Estates, 2001a). However, contracting out of cleaning services in hospitals has continued (Unison, 2005a). This was highlighted in 2002 by the Minister of Health who confirmed that NHS Trusts will continue to "award contracts on best value taking into account operational and economic issues" (Hansard, 2002).

Since 2000, there have been increasing concerns over cleaning standards (Unison, 2005a), which has led to the publication of a number of initiatives including the *Clean Hospitals Programme* (NHS Estates, 2001a), *National Standards of Cleanliness in the NHS* (NHS Estates, 2001b), *National Standards of Cleanliness for NHS Trusts in Wales* (Welsh Assembly Government 2003), the *Healthcare Facilities Cleaning Manual* (NHS Estates, 2004b) and the *Revised Guidance on Contracting for Cleaning* (NHS Estates, 2004a). The revised guidance on contracting for cleaning has been designed to offer best practice on evaluating and awarding contracts so that quality is considered alongside price (NHS Estates, 2004a). This guidance has, however, been the subject of criticism from Unison (2005b) as it was been developed without consultation from anyone who actually carries out cleaning. Unison has also expressed concern that the guidance fails to address "the ongoing problem of under-resourcing of hospital cleaning which has prevailed since the competitive tendering regime began to oblige trusts to accept the lowest price tender regardless of quality" (Unison 2005b). It is reasonable to point out that the remit of this guidance was to provide "the standards relating to hospital cleaning that all NHS trusts should follow as a minimum" (NHS Estates, 2004a), rather than to challenge the current arrangements for hospital cleaning. Nonetheless, the comments from Unison highlight the depth of feeling that currently exists relating to this subject.

The incidence and costs of hospital acquired infections in the NHS

A *hospital acquired infection* is an infection that is neither present nor incubating when a patient is admitted to hospital which normally manifests itself more than 48 hours after admission to hospital (National Audit Office, 2004). To avoid confusion, however, it is important to be aware of the term *healthcare associated infection*, which is an infection acquired via the provision of healthcare in either a hospital or community setting (Department of Health, 2003). It can be seen that the latter term has a broader remit and also covers the provision of healthcare by primary care trusts.

Unison (2005a) discovered that it is not a straightforward exercise to track the incidence of hospital acquired infection rates that replicates the findings of "an Organisation with a Memory" (Department of Health, 2000a). This report found that data are not analysed to identify patterns or trends, there is no consensus on what to report and there are no proper linkages between reporting systems. In addition, the NAO (National Audit Office, 2000) discovered there was no requirement for NHS Trusts to publish data on hospital acquired infection, and such data that had been published was not comparable. Despite these challenges, the NAO managed to conclude that at any one time nine per cent of patients had an infection that had been acquired during their hospital stay, which equated to at least 100,000 infections a year (National Audit Office, 2000). These data are based upon two national prevalence surveys that were published in the United Kingdom in 1981 and 1996. However, these are not recent data; this highlights the problems encountered by the NAO when collating figures for its report in 2000. The same report also found that 5,000 patient deaths each year might be primarily attributable to hospital acquired infection. These figures were based upon extrapolation of data acquired from the United Sates in the mid 1980s, although the difficulties of undertaking such an analysis were acknowledged (National Audit Office, 2000).

The DH has subsequently taken action to try and address the lack of meaningful data relating to the incidence of hospital acquired infection. For example, in 2001 the DH began a mandatory surveillance scheme of methicillin resistant staphylococcus aureus (MRSA) and from April 2005, NHS acute trusts have been set the target of reducing MRSA bacteraemia (blood stream infection) rates year on year (Department of Health, 2004a). The results of the MRSA bacteraemia surveillance scheme are available on the DH webpages (www.doh.gov.uk) and highlight the numbers and rates of MRSA bacteraemias per 1000 bed days in NHS acute trusts from April 2001 until September 2004.

Although it is not easy to make comparisons from the data because of their layout, and there are several cautionary notes, the DH does provide information (Department of Health, 2005) that shows the total number of MRSA bacteraemias in England between April and September 2004 fell to 3519. The corresponding figure for the same time period in the previous three years was 3598 (2001), 3574 (2002) and 3744 (2003). This information is shown in figure 8.1. Conversely, the number of MRSA bacteraemias in the first three complete years of the mandatory recording system rose from 7249 in 2001/02 to 7373 in 2002/03 and 7684 in 2003/04. This information is shown in figure 8.2. Unfortunately, the DH offers no assistance with the interpretation of these results to explain why the overall figures vary so significantly when compared by a six month period and then by a year.

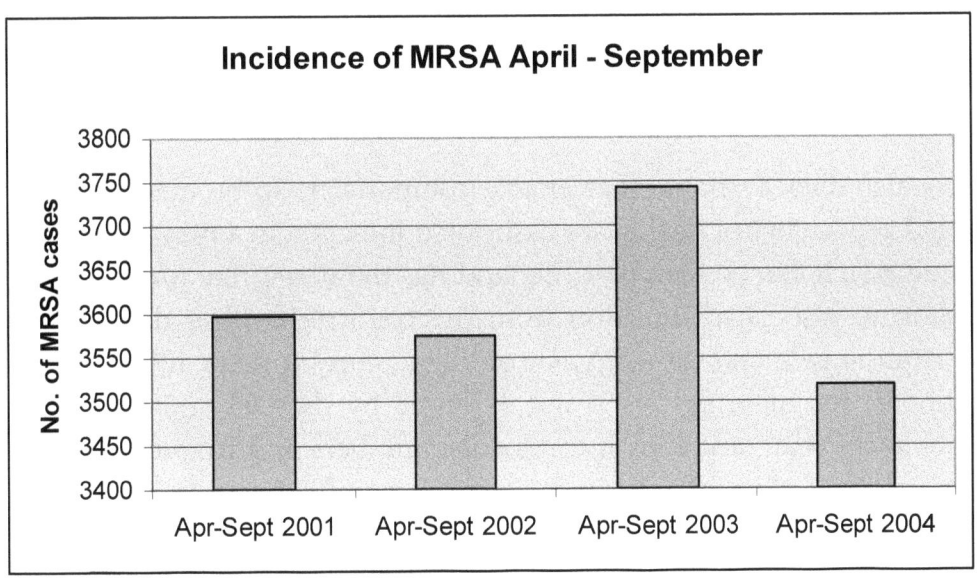

Figure 8.1 – Incidence of MRSA: April to September 2004 (Source Department of Health, 2005)

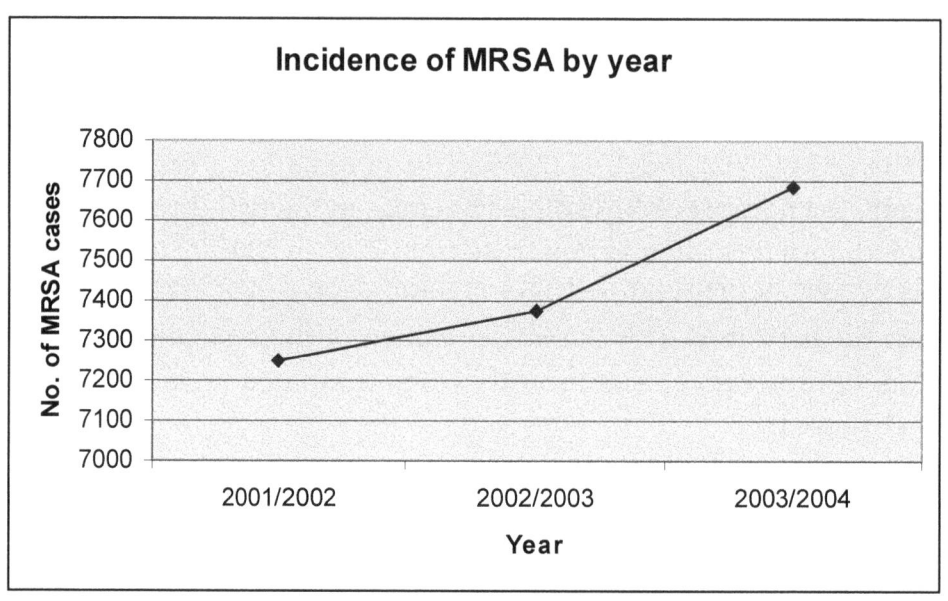

Figure 8.2 - Incidence of MRSA by year (Source Department of Health, 2005)

There is also data available that enable comparisons to be made between the estimated prevalence of healthcare associated infection in a number of countries (see table 9.1). It can be seen that England has the worst rate for the prevalence of healthcare associated infection from the countries whose data is available (apart from France and the USA whose range encompasses 10%). Once again, the rationale for these figures is not explored by the DH, although the NAO (2004) believes that variations in protocols, numbers and frequency of hospital participation actually make direct comparison unreliable.

Country	Prevalence of healthcare associated infection
Australia	6%
Denmark	8%
England	9%
France	6 – 10%
Netherlands	7%
Norway	7%
Spain	8%
USA	5 – 10%

Table 9.1 – Prevalence of healthcare associated infection.
(Source: Department of Health, 2003)

Data are also collected by the NHS to show trends in surgical site infection in the national nosocomial infection surveillance system. A review of information collected from over 60,000 operations over a five year period from 1997 – 2001 showed that while 12 per cent of hospitals reduced their rates of infection in this area, 2.5 per cent had increased rates and the vast majority had shown no improvement (Department of Health, 2003). However, it must be acknowledged that, despite the data that are now accessible on MRSA bacteraemia and surgical infection rates, there is little other comparable data currently available. For example, national data are not collated for the two commonest sites of infection – the urinary tract and lung infections - which constitute 23 percent and 22 percent respectively of all known infections (Emmerson et al, 1996). This makes it difficult to quantify with any certainty whether there have been any real changes in NHS Trust's infection rates (National Audit Office, 2004).

Whilst the available information relating to the incidence of hospital acquired infection is limited and patchy, the knowledge about its costs is even more limited. The DH (2003) acknowledges that the costs of treatment are difficult to measure with certainty but accept that they are "high". It has been estimated by the NAO (2000) to be as much as £1 billion each year, and 15 percent of that figure could be saved by better application of good infection control practice. This would release £150 million for alternative NHS use (National Audit Office, 2000). The DH (2003) cites data from the United States relating to the costs of infection control which have demonstrated that the estimated annual cost of infection control is US$800,000, whilst a 32 percent reduction in baseline rates of hospital acquired infection would lead to savings of US$2,400,000.

Only 11 per cent of NHS trusts have undertaken similar economic evaluations based on the principles of the 2000 NAO report; and whilst these showed a variety of results, they all demonstrated the significant financial burden of hospital acquired infection (National Audit Office, 2004). Consequently, the NAO (2004) notes that as the availability of cost information has not improved, the data from their 2000 report remain the "best estimate of the overall cost to the NHS currently available."

Assessment of hospital cleanliness in the NHS

In 2000, following the publication of the NHS plan (Department of Health, 2000b), independent Patient Environment Action Teams (PEATs) were established to review standards of cleanliness at acute hospitals from a patient's perspective (Department of Health, 2004a). This is significant since cleanliness and infection control are believed to be closely linked. PEATs consist of NHS staff, patients, patient representatives and members from the general public (Unison, 2005a).

When the assessment process was reviewed in 2004, a new scoring mechanism was introduced which moved away from the three-point 'traffic light' method to a five-point scale to enable clearer distinctions to be drawn (NHS Estates, 2005). The results of these assessments have shown steady improvement since they commenced, although the new scoring system does make comparisons with previous years more complicated (see figure 8.3, tables 8.2 and 8.3). It must also be highlighted that the reviews do not focus solely on cleanliness as they also include areas such as, food and food service, parking, privacy and dignity – as pointed out by Unison (2005a), but not the DH (2004a).

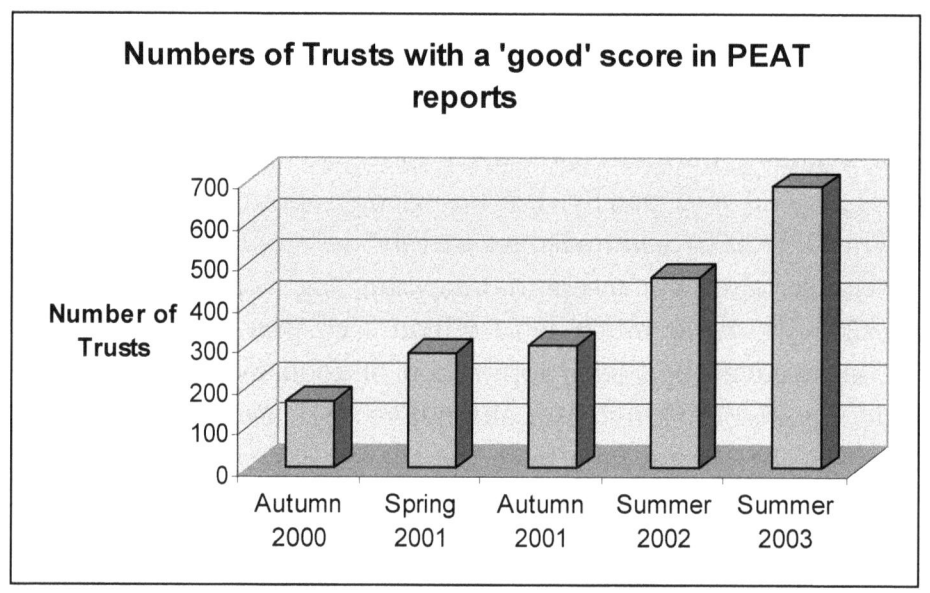

Figure 8.3 – Percentage of Trusts achieving a good standard of cleanliness in PEAT reports (Source: Department of Health, 2004a)

	Red (poor)	Yellow (acceptable)	Green (good)
Autumn 2000	253 (35.5%)	297 (41.7%)	163 (22.3%)
Spring 2001	42 (6.1%)	368 (53.4%)	279 (40.5%)
Autumn 2001	0 (0%)	387 (56.3%)	300 (43.7%)
2002	0 (0%)	317 (40%)	464 (60%)
2003	0 (0%)	186 (21.3%)	686 (78.7%)

Table 8.2 - Results of PEAT reviews 2000 – 2003
(Source: NHS Estates, 2005)

2004	Excellent	Good	Acceptable	Poor	Unacceptable
Number of hospitals	118 (10%)	456 (38.5%)	583 (49%)	24 (2%)	3 (0.5%)

Table 8.3 – Results of PEAT reviews 2004
(Source: NHS Estates, 2005)

Factors influencing the rates of hospital acquired infections

a) Contracting out of cleaning services

As previously discussed, the number of cleaners has fallen since the introduction of contract cleaning, but the impact of this action on the rates of hospital acquired infections is less easy to determine. The impact of cleaning itself on infection in hospital continues to be debated. Midgley (2001) reports that there is no hard evidence linking dirty hospitals with the spread of hospital acquired infections, although (s)he recognises that common sense would suggest that the two are related.

In contrast, the Auditor General for Wales (2003) is adamant that effective cleaning is vital to maintain a healthy and safe hospital environment, since a clean hospital helps to limit the risk of infection. Rampling et al (2001) support this view when they describe how increasing the number of cleaning hours by fifty seven per week controlled an outbreak of MRSA and reduced the incidence from sixty nine to three patients. Dancer (1999) also believes that hospital cleaning is a method of controlling hospital-acquired infection; and considers contracting out of hospital cleaning services has contributed to falling standards of cleanliness. Unison (2005a) reinforce this view and cite specific problems relating to drawing up contracts, lack of flexibility, lack of trust and monitoring, difficulties in imposing sanctions, separation of cleaning services, damage to the public sector ethos and problems with health and safety training, which have all had an impact on cleaning standards.

Unison (2005a) also report that contracting out of cleaning services brings with it low pay, poor conditions, an intensification of work and a decline in job satisfaction, which lead to difficulties in recruiting and retaining staff and high levels of sickness, which again have repercussions for standards of cleaning (Unison, 2005a). The issues relating to low morale and reduced job satisfaction are addressed in part by the Matron's Charter which stresses the importance of ensuring cleaning staff are seen as an integral part of the ward team, suggesting that they should be invited to nights out and have their photographs displayed with other members of the ward team (Department of Health, 2004b).

Conversely, cleaning contractors insist there is no link between dirty wards, outsourcing and rates of infection, but because there is a lack of definitive data, this perspective is difficult to substantiate (Gosling, 2004). Gosling (2004) quotes the Director-general of the Cleaning and Support Services Association (CSSA), who believes the quality of cleaning relates to the resources allocated rather than whether the provider operates in or out of house, estimating that expenditure on cleaning is generally about 20 percent less than is needed. He states the way forward is not to look at who is providing the cleaning service, but to look at the cleaning specification; and believes the review of standard specifications and frequencies of cleaning recently published by NHS Estates (2004b) will be beneficial in raising standards of hygiene and reducing infection rates.

This view is shared by the Auditor General for Wales (2003) who found significant variations in the cleaning approaches adopted by trusts irrespective of whether the cleaners were trust based or contracted out. This was felt to be as a result of poorly defined cleaning specifications which did not help staff achieve the desired outcome – a clean hospital.

Unison (2005a) highlights two distinct criteria that relate to the effects that the contracting out of cleaning services has had on standards of cleanliness and subsequent levels of infection. The first centres on the effects of tendering (whether or not the service is contracted out) and the second applies specifically to the contracting out process. However, there has been little research undertaken in either area and Unison recognises that there is a lack of "comparative longitudinal data" that compares links between contract cleaners and the trends of increasing infection rates. Unison blames the DH for this lack of information despite acknowledgement from the Secretary of State that "compulsive competitive tendering has gone – because it failed to raise standards" (NHS Estates, 2001b).

Unison feel that real progress can be made by taking the simple and effective step of bringing hospital cleaning back in-house and providing sufficient resources for decent pay and staffing levels. There is, however, limited evidence for the impact of this strategy on rates of hospital infections.

A brief comparison between a small number of trusts that have contract cleaners with trusts that employ their own cleaners shows no real correlation between the type of cleaners in place and incidence levels of MRSA (see table 8.4).

Name of Trust	Apr 01 – Sept 01	Oct 01 – Mar 02	Apr 02 - Sept 02	Oct 02 – Mar 03
Hinchingbrook (General)	**4**	**8**	**8**	**17**
Rotherham (General)	*9*	*13*	*10*	*4*
Royal Cornwall (General)	*37*	*18*	*17*	*34*
Nuffield Orthopaedic Centre (Specialist)	**0**	**1**	**2**	**1**
Christie Hospital (Specialist)	*7*	*5*	*6*	*3*

Table 8.4 – Number of MRSA cases reported by NHS Trusts. Trusts in bold font have contract cleaners. Trusts in italic font have NHS employed cleaners. (Source: Department of Health, 2005)

b) <u>education and training</u>

The importance of education and training in infection control for staff in the healthcare setting is reinforced by the DH, the NAO and the NHS Litigation Authority (NHSLA). The NHSLA is a special health authority which has an active risk management programme to help raise standards of care in the NHS (NHSLA 2004a). This programme is managed through a range of standards and assessments, one of which is the Clinical Negligence Scheme for Trusts (CNST) which assesses an organisation's approach to a variety of risks including infection control and staff training (NHSLA, 2004a).

The NAO (2000) strongly recommends that NHS trusts should ensure all staff are targeted through induction training and that staff who have day to day contact with patients are kept up to date on good infection control practice. Similarly, one of the 10 commitments from the Matron's Charter is to ensure that all staff working in healthcare receive education in infection control (Department of Health, 2004b). Unfortunately, the increase in rates of hospital acquired infection could be attributed in part to a lack of training since there are currently some ten percent of infection control teams who do not provide nurses and health care assistants with induction training about infection control; and less than two thirds provide annual updates. Most teams do not provide any infection control training to senior doctors (National Audit Office, 2000)

Evidence relating to the effectiveness of training programmes is provided by Akid (2004), who describes the development of a training programme to reduce infections caused by poor aseptic techniques used for clinical procedures, including intravenous injections. As a result of its success in improving aseptic techniques and reducing the rates of MRSA, the technique is, at the time of writing, being rolled out to the remaining seven hospitals in the London Trust.

c) <u>handwashing</u>

It is believed that effective hand hygiene is possibly the most effective method of preventing hospital acquired infection (Teare, 1999). The DH (2003) also address the importance of handwashing by recognising that "healthcare workers are a major route through which patients become infected." In addition, the CNST standards (NHSLA, 2004b) highlight the importance of hand hygiene by assessing hand washing training and requesting evidence that staff attend these sessions.

Reviews of research support the view that hand washing has a significant impact on the rates of hospital acquired infections. For example, Stone et al (2001) reviewed nine studies (three randomised controlled trials, five controlled trials, and one multiple crossover trial) and found that hand hygiene led to major reductions in infection related outcomes across a wide range of clinical settings. They felt the effect was so great that if "hand hygiene" were a new drug it would be accepted without question (Stone et al, 2001). Similarly, the Royal College of Nursing (RCN) recognise that handwashing is the single most important activity for reducing the spread of disease, whilst highlighting evidence that many healthcare professionals do not use the correct technique (RCN, 2003). This view is supported by Roberts (2002) who reviewed many observational studies which showed low rates of handwashing, especially amongst doctors. Reasons for this are varied and include lack of time and a paucity of hand hygiene agents (Roberts, 2002 and Department of Health, 2003). A number of local initiatives have been implemented to encourage hand washing, which include red arrows on the floor pointing to wash hand basins to guide staff and visitors to wash their hands and the use of trust-wide screen savers with hand washing instructions (Pugh, 2005).

To support these ideas, the National Patient Safety Agency (NPSA) developed the 'cleanyourhands campaign' to address the "unacceptable low levels of hand hygiene compliance amongst NHS staff" (NPSA, 2004). The campaign includes the implementation of near-patient alcohol hand rubs, a series of posters and supporting marketing materials, patient information and the use of ward-based champions to drive the initiative forward. A pilot study was initially undertaken to evaluate the effectiveness of the campaign, and the results showed that the practice of hand hygiene moved from low priority to a core element of daily practice that could easily be achieved (National Patient Safety Agency, 2004).

d) <u>audit and surveillance of hospital acquired infection</u>

The first action area from 'Winning Ways' (Department of Health, 2003) highlights the importance of collating high quality information on healthcare associated infection since this is essential to "tracking progress, investigating underlying causes and instituting prevention and control measures."

The NAO (2000) found that infection control was largely reactive in nature and, whilst only 50 percent of infection control teams included audit in their annual infection control programme, 81 percent of infection control teams had not audited their own activities. The NAO also discovered a lack of detailed surveillance information, which meant that many health authorities do not have the data they need to assess NHS Trusts' performance in improving infection control, nor are hospitals able to prioritise their resources for dealing with hospital acquired infection (National Audit Office, 2000).

Surveillance is seen as essential since data collection, analysis and feedback of results to clinicians is central to detecting infections, dealing with them and ultimately reducing infection rates (National Audit Office, 2000). Disappointingly, the NAO (2004) reports that "there is still no comprehensive mandatory surveillance scheme."

e) <u>strategic management of hospital acquired infection</u>

The Auditor General for Wales (2003) found that cleaning was not a priority for trust boards, and felt this was one factor in the increasing rates of hospital acquired infection. This finding reflects the NAO (2000) report which found that, despite chief executives having overall responsibility for ensuring the provision of effective infection control arrangements, 58 percent of them never received reports on resources spent on infection control and less than half received reports on rates or numbers of hospital acquired infections. The NAO also found that a quarter of service agreements between NHS trusts and their health authorities did not cover the provision of infection control services).

A number of initiatives have been implemented to raise the profile of hospital acquired infection and improve its prevention and control (National Audit Office, 2000). These include an annual review of infection control arrangements, as required by CNST (NHSLA 2004b), and the designation of a Director of Infection Prevention and Control (DIPC) who reports directly to the Chief Executive and Trust Board (Department of Health, 2003).

Significant changes have been made to the strategic management of infection control, which is now seen as an NHS priority (National Audit Office, 2004). And there are a number of regulatory and other bodies that now exist that have responsibilities for infection control though external performance monitoring (see appendix 8.1).

f) <u>funding arrangements for infection control</u>

It is important for NHS Trusts to make adequate funding arrangements for infection control so that rates for hospital acquired infections may be reduced. At the current time, there is a mismatch between what is expected of the infection control team and the staffing and resources allocated to them (National Audit Office, 2000). This situation is made worse by the fact there are no guidelines on infection control staffing from the DH. Consequently, there are wide and unacceptable variations in the ratio of infection control nurses to beds, which impact on the rates of hospital acquired infections (National Audit Office, 2000). The NAO (2004) recognise that some trusts are attempting to address this issue, but discovered that whilst two-thirds of chief executives approved changes to infection control staffing resources, fewer than half approved changes to the non-pay budget and, in some trusts, the infection control budget had actually decreased (National Audit Office, 2004).

g) <u>new roles and responsibilities for healthcare staff</u>

There have been two new key roles introduced recently – the modern matron and the ward housekeeper - which have the potential to impact significantly on the rates of hospital acquired infection.

The role of the modern matron was first highlighted in the NHS Plan (Department of Health, 2000b) and includes a responsibility to lead clinical teams in the prevention of healthcare associated infections (Department of Health, 2001) The NAO (2004) emphasises that infection control is an intrinsic part of the role of the modern matron. Since 1999, more than 3,000 modern matrons have been appointed in the NHS (Department of Health, 2004a) so that all acute trusts now have modern matrons (Pugh 2005). Hill and Hadfield (2005) describe the difference a modern matron made by introducing collaborative working and infection control audits of the environment and practice, which were subsequently turned into action plans leading to changes in practice.

The ward housekeeper's role was also introduced in the NHS Plan (Department of Health, 2000b) and has 11 patient focused standards, one of which is cleanliness. By September 2004, the NHS had ward housekeepers in 53 per cent of all hospitals, rising to 70 per cent in the larger hospitals where the majority of patients receive treatment (Department of Health, 2004c). Reports from NHS Estates (2004c) describe how the housekeepers' influence on cleanliness has been "dramatic." East Somerset Hospitals NHS Trust found that compliance with cleanliness standards in four wards on two floors rose from 82 percent and 60 percent to 97 percent and 96 percent respectively, whilst housekeepers in Oxford have lifted cleaning standards scores by up to 30 percent on the 24 wards in which they are based. Further evidence relating to the benefits of the ward housekeeper service is provided by St George's Healthcare NHS Trust where, before the service was provided, there was 100 percent dissatisfaction with the cleanliness of the ward environment – compared with a current 80 percent satisfaction (NHS Estates, 2004c).

h) <u>antibiotic prescribing</u>

Pugh (2005) reports that the over-prescription of antibiotics is partly to blame for the upsurge in rates of hospital acquired infections, whilst the DH (2003) recognises that escalating antibiotic resistance is making many infections difficult to treat. The DH (2003) is keen to promote the 'prudent use of antibiotics' since "indiscriminate and inappropriate use of antibiotics to treat infection within a clinical service promotes the emergence of antibiotic resistant organisms and the 'super-bug' strains." However, this desire may prove difficult to consummate as illustrated by research undertaken by Wester et al (2002) who surveyed 490 doctors to assess their attitudes about the importance of antibiotic resistance. They discovered that, although most physicians (97%) viewed antibiotic resistance as a serious national problem, only 60% favoured restricting use of broad-spectrum antibiotics. This raises concerns that disparities in doctors' beliefs and attitudes may compromise efforts to improve antibiotic prescribing and infection control practices (Wester et al, 2002).

i) <u>hospital design</u>

The DH (2003) recognises that "the risks of healthcare associated infection are greatly increased by an absence of suitable facilities to isolate infected patients." Research is being undertaken to determine how many single rooms are required to assist in the implementation of infection control, and new hospitals being built have more single rooms than ever before (Department of Health, 2004a). This action reflects the belief that the control of infection needs to be "designed into" hospitals (Department of Health, 2004a). It is also crucial that any new building or refurbishment must ensure all areas are accessible for cleaning, and provide enough storage to prevent clutter (Department of Health, 2004b). There is little doubt that poor bed space design encourages MRSA and other forms of hospital acquired infection to flourish (Lindsay 2004).

Infection control teams must also be involved with any construction or renovation project, since construction dust and exposure to previously enclosed areas of a building can lead to more infections (McLaughlin and McMacken, 1999).

O'Connell and Humphreys (2000) report that a number of professional and scientific bodies in the UK, the USA and Europe have published guidelines on the design and layout of ICUs. They all emphasise the importance of adequate isolation facilities to help minimise infection in this high-risk area.

j) <u>medical devices and equipment management</u>

According to the DH (2003), "The two strongest risk factors linked with healthcare associated infection are the degree of underlying illness and the use of medical devices." The important role of medical devices is emphasised by the 80% of urinary infections that are traced to indwelling catheters and the fact that over 60% of blood infections are introduced by intravenous feeding lines or similar devices (Department of Health, 2003).

As inadequate decontamination has frequently been associated with outbreaks of infection in hospital, it is vital that reusable equipment is scrupulously decontaminated between each patient. To ensure that control of infection is maintained at a high level, all health care staff must be aware of the implications of safe decontamination (RCN, 2003). To highlight the importance of this issue, the second action area from 'Winning Ways' (Department of Health, 2003) relates to reducing the risk of infection from the use of "catheters, tubes, cannulae, instruments and other devices." The importance of training and competence in using aseptic techniques is highlighted together with the realisation that such devices must be kept in place for the minimum time necessary (Department of Health, 2003).

Due to the word limitations of this chapter, it has not been possible to discuss other factors that may impact on the incidence of hospital acquired infections. For example, bed shortages and high bed occupancy mean some patients are nursed on inappropriate wards and then moved to the correct ward when a bed becomes available, thereby increasing the risk of cross infection (Pugh, 2005). In addition, the prevalence of open visiting times, the increasing lack of restrictions on the numbers of visitors and allowing visitors to stay overnight on wards when a patient is unwell are factors attributing to increasing infection rates (Maternity Visiting Group, 2005).

It is also important to consider that, whilst improvements in medical care are prolonging life, they are also creating a high proportion of patients with weakened immune systems who are more susceptible to infection (Spencer, 2004).

Finally, there is the vast subject of food and kitchen hygiene which has the potential to cause havoc within a healthcare environment. A study undertaken by Griffiths et al (2000) found that sites most likely to fail ward-based cleanliness assessments were the toilet and kitchen - areas which are frequently implicated in the spread of infectious intestinal diseases.

Conclusions

This chapter has provided evidence that, despite a substantial lack of available credible data, the rates of hospital acquired infections in England and the UK are increasing and are of significant concern. The costs of hospital acquired infections are also serious, not only in monetary terms but also in terms of personal cost to the patient and their family.

It has not been possible, however, to demonstrate that increased rates of hospital acquired infections have been as a result of the introduction of contracting out of cleaning services in the NHS since 1983. There is little doubt that this action reduced the numbers of cleaners in the NHS and subdued the perceived importance of cleanliness in hospitals. Consequently, it would be reasonable to assume that a link between the rise in hospital acquired infections and hospital cleaning is possible, but this has not been proven (Unison, 2005a).

Even if data had been available to prove such a link, it would need to be recognised, as this assignment has demonstrated, that contracting out of cleaning services is clearly not the only reason for the spread of hospital acquired infections (Unison, 2005a). As the DH (2004a) reinforces, cleanliness contributes to infection control, but preventing infections requires more than cleanliness.

No single factor explains the growth in the number of patients who acquire infections during the course of their treatment by the NHS or other healthcare systems around the world (Department of Health, 2003). As has been discussed, there are a multitude of factors that affect the incidence of hospital acquired infections and there now needs to be a shared and collective responsibility to addressing this problem.

To help reduce the incidence of hospital acquired infection, the message for hospital staff must be that "infection control is everybody's business and there's something that everybody can do to prevent it" (Pugh, 2005). This is reiterated in the first commitment of the matron's charter which states "keeping the NHS clean is everybody's responsibility" (Department of Health, 2004b).

Appendix 8.1 - Independent bodies with responsibilities for infection control (Source: National Audit Office, 2004)

The NHS Litigation Authority

Handles the Clinical Negligence Scheme for Trusts, which established standards in 1999 to provide a framework for clinical risk management, including infection control. Assesses trusts against these standards.

The National Patient Safety Agency

Formed in 2001. Main role is to establish and manage a national reporting system to learn from adverse patient incidents, including hospital acquired infections. They also initiate preventative measures to help reduce unintended harm to patients, including the "cleanyourhands" campaign.

The NHS Purchasing and Supply Agency

Established in 2000 and is responsible for trusts' purchasing policies. Introduced high quality paper towels and is supporting the "cleanyourhands" campaign by developing a range of alcohol hand rubs and containers that meet the unique requirements of the NHS.

NHS Estates
[**Editor's note**: NHS Estates no longer exists. Refer to the Department of Health website for latest information on infection control issues at www.dh.gov.uk].

Published Infection Control in the Built Environment in 2001, providing guidance on the planning, design and maintenance of the healthcare buildings and equipment. Also produced National Standards of Cleanliness. Patient Environment Action Teams (PEATs) undertake reviews on aspects of the patient's environment.

Health and Safety Executive

Carries out planned inspections of health and safety standards in healthcare premises, and may also become involved in investigations following cases of occupational disease or serious incidents following patient infections, although this rarely occurs in practice.

Strategic Health Authorities (replaced by Regional Authorities)

Monitor performance of trusts and are accountable for delivery of targets. Infection control did not feature in these until June 2003. Also reviewed compliance with Controls Assurance Standards.

Commission for Health Improvement (replaced by the Healthcare Commission from 1/4/2004)

Established in 1999. Reviewed clinical governance arrangement in trusts, and regularly reviewed infection control arrangements. Published performance ratings for NHS Trusts for the first time in 2003. MRSA bacteraemia improvement scores and infection control standard scores were included for the first time in 2002/2003.

Medicines and Healthcare related products Regulatory Agency

Formed from the Medical Devices Agency and the Medicines Control Agency in 2003. Investigates adverse incidents related to medical devices including those arising from decontamination problem. Issues device bulletins as a result of experience gained from adverse incident investigations.

The Health Protection Agency

Formed in 2003 and dedicated to protecting people's health and reducing the impact of infectious diseases (taking over from the former Public Health Laboratory Service), chemical hazards, poisons and radiation hazards. A key responsibility is monitoring and helping to manage outbreaks of hospital acquired infection. The Department of Health also had a service level agreement with the Public Health Laboratory Service which was transferred to the HPA, to develop surveillance of infection rates.

National Institute for Health and Clinical Excellence

Established in 1999 to provide patients, health professionals and the public with authoritative, robust and reliable guidance on current "best practice". Published guidelines on infection control in primary and community care in 2003.

References

Akid M (2004). No space for invaders. *NHS Magazine* February 16-17.

Auditor General for Wales (2003) *The Management and Delivery of Hospital Cleaning Services in Wales.*
www.wao.gov.uk/assets/englishdocuments/Management_and_Delivery_of_Hospital_Cleaning_Services_agw_2003.pdf

Conservative Party (1983). *Conservative Party General Election Manifesto.*
www.conservative-party.net/manifestos/1983/1983-conservative-manifesto.shtml

Dancer S (1999). Mopping up hospital infection. *Journal of Hospital Infection* 43 (2) 85-100.

Department of Health (2000a). *An Organisation with a Memory*. HMSO, London.

Department of Health (2000b). *The NHS Plan: a plan for investment, a plan for reform*. HMSO, London.

Department of Health (2001). *HSC 2001/010: Implementing the NHS Plan: modern matrons: strengthening the role of ward sisters and introducing senior sisters.* HMSO, London.

Department of Health (2003). *Winning Ways*. HMSO, London.

Department of Health (2004a). *Towards Cleaner Hospitals and Lower Rates of Infection*. HMSO, London.

Department of Health (2004b). *A Matron's Charter: An Action Plan for Cleaner Hospitals*. HMSO, London.

Department of Health (2004c). *Government hits target of housekeepers in over half of all NHS hospitals.*
www.dh.gov.uk/PublicationsAndStatistics/PressReleases/PressReleasesNotices/fs/en?CONTENT_ID=4099182&chk=nsZ9qZ

Department of Health (2005). *MRSA surveillance system results.*
www.dh.gov.uk/PublicationsAndStatistics/Publications/PublicationsStatistics/PublicationsStatisticsArticle/fs/en?CONTENT_ID=4085951&chk=HBt2QD

Domberger S, Meadowcraft S and Thompson D (1987). The impact of competitive tendering on the costs of hospital domestic services. *Fiscal Studies* Vol 8:4.

EMIRE (2005). *Fair Wages Resolution*. www.eurofound.eu.int/emire/UNITED%20KINGDOM/FAIRWAGESRESOLUTION-EN.html

Emmerson A, Enstone J, Griffin M, Kelsey M and Smyth E (1996). The second national prevalence survey of infection in hospitals – overview of results. *Journal of Hospital Infection* 32. 175-190.

Gosling P. (2004) Source of infection? *People Management* (10) 18. 42-44.

Griffiths C, Cooper R, Gilmor, J, Davies C and Lewis M (2000). An evaluation of hospital cleaning regimes and standards. *Journal of Hospital Infection* 45 (1) 19-28.

Hansard (2000). Cleaning and Hygiene Services. *Written Answers* 12 April, col.211.

Hansard (2002). Hospital Cleaning Contracts. *Written Answers* 17 April, col.1034.

Hill D and Hadfield J (2005). The role of modern matrons in infection control. *Nursing Standard* Feb 16, 19 (23). 42-44

Jones N (2004). The legacy of King Arthur. *BBC News* 10th March. http://news.bbc.co.uk/2/hi/uk_news/3499611.stm

Lindsay S (2004). Healthcare associated infections: word is spreading. *NHS Litigation Authority Review* 31 (6-9). www.nhsla.com

Maternity Visiting Group (2005.) Shortcuts. *Smartpractice* February Issue 11. 5. www.gmsha.nhs.uk/smart/

McLaughlin S and McMacken D (1999). Germ Warfare! *Health Facilities Management* 12 (1). 12-16.

Midgley S (2001). Looking Good. *NHS Magazine* April, 2 (4). 19, 21.

Milne R G (1993). Contractors experience of compulsive competitive tendering: a case study of contract cleaners in the NHS. *Public Administration* Autumn, (71). 301-321.

Milne R G and McGee M A (1992). Compulsory Competitive Tendering in the NHS: a new look at some old estimates. *Fiscal Studies* (13) 3. 109-130.

Milne R G and Wright R E (2004). Competition and costs: evidence from competitive tendering in the Scottish National Health Service. *Scottish Journal of Political Economy* (51)1. 1-23.

National Audit Office (2000). *The Management and Control of Hospital Acquired Infection in Acute NHS Trusts in England.* The Stationery Office, London.

National Audit Office (2004). *Improving patient care by reducing the risk of hospital acquired infection: A progress report.* The Stationery Office, London.

NHS Estates (2001a). *The NHS Plan – Clean Hospitals.* Department of Health Publications, London.

NHS Estates (2001b). *National Standards of Cleanliness for the NHS.* Department of Health Publications, London.

NHS Estates (2004a). *Revised Guidance on Contracting for Cleaning.* Department of Health Publications, London.

NHS Estates (2004b). *The Healthcare Facilities Cleaning Manual.* Department of Health Publications, London.

NHS Estates (2004c). *Ward Housekeepers*. Department of Health Publications, London.

NHS Estates (2005). *Clean Hospitals & PEAT.*
http://patientexperience.nhsestates.gov.uk/clean_hospitals/ch_content/national_results/introduction.asp

NHS Litigation Authority (2004a). *About the NHS Litigation Authority.* www.nhsla.com/home.

NHS Litigation Authority (2004b). *Clinical Negligence Scheme for Trusts – Clinical Risk Management Standards.* www.nhsla.com

National Patient Safety Agency (2004). *Achieving our aims. Evaluating the results of the pilot CleanyourHands campaign.*
www.npsa.nhs.uk/site/media/documents/692_final_evaluation.pdf

O'Connell N and Humphreys H (2000). Intensive care unit design and environmental factors in the acquisition of infection. *Journal of Hospital Infection* 45 (4) 255 - 262.

Pugh R (2005) All cleaned up? *Smartpractice* February Issue 11. 4 - 5. www.gmsha.nhs.uk/smart/

Rampling A, Wiseman S, Davis L, Hyett A, Walbridge A, Payne G and Cornaby A (2001). Evidence that hospital hygiene is important in the control of methicillin-resistant Staphylococcus aureus. *Journal of Hospital Infection* 49 (2) 109 – 116.

RCN (2003). *Good practice in infection control – guidance for nursing staff.* www.rcn.org.uk/resources/mrsa/downloads/Wipe_it_out-Good_practice_in_infection_prevention_and_control.pdf

Revill J (2005). Scandal of staff cuts in filthy wards. *The Observer* January 9th.

Roberts G (2002). *Risk Management in Healthcare.* 2nd ed. Witherby & Co. Ltd, London.

Spencer A (2004) Infection control. *NHS Magazine* September. 18-19.

Stone S, Teare E and Cookson B (2001). The evidence for hand hygiene. *The Lancet* (357). 479 – 480.

Teare E (1999). Handwashing. A Modest Measure with big effects. *British Medical Journal* (318). 686.

Unison (2003). *NHS Market Report.* April. 1-8. www.unison.org.uk/acrobat/B828.pdf

Unison (2005a). *Hospital contract cleaning and infection control.* UNISON, London.

Unison (2005b). *Cleaners' voices / Interviews with hospital cleaning staff.* UNISON, London.

Welsh Assembly Government (2003). *The National Standards of Cleanliness for NHS Trusts in Wales.* www.wales.gov.uk/subihealth/content/keypubs/circulars/whc-2003-059-pt3-e.pdf

Wester C, Durairaj L, Evans A, Schwartz D, Husain S and Martinez E (2002). Antibiotic Resistance: A Survey of Physician Perceptions. *Archives of Internal Medicine* 162 (19). 2210 – 2216.

This page is intentionally blank.

9

Reducing the frequency and impact of needlestick injuries involving healthcare staff

JAYNE HARTLEY

Introduction

The intention of this chapter is to review the effectiveness of systems that have been put in place to reduce the cause and effects of needlestick injuries to staff working in healthcare.

Before commencing the review, the key terminology to be used will be clarified. This will focus on the terms 'needlestick injury', 'cause' and 'effect'. This will be followed by a detailed exploration of the incidence of needlestick injuries in the United Kingdom (UK), Europe and the United States of America (USA). It will be seen that there are significant numbers of needlestick injuries reported each year which mostly affect nurses, junior doctors and ancillary staff.

The causes of needlestick injuries will then be considered. This will include a review of devices that are responsible for the majority of needlestick injuries and consideration of those circumstances which might lead to such an injury. The systems that have been, or are being, developed in an attempt to reduce the incidence of needlestick injuries will then be highlighted. This will include the training and education of healthcare workers, the application of universal precautions and the use of safer needle devices. The success of these systems will also be debated.

The effects of needlestick injuries upon healthcare workers will then be considered. This section will include consideration of the physical and psychological consequences of the injury, possible ill health retirement and claims for compensation. It will also discuss three of the main blood-borne pathogens that staff are most at risk of developing following exposure as a result of a needlestick injury. These are Hepatitis B Virus (HBV), Hepatitis C Virus (HCV) and Human Immunodeficiency Virus (HIV). The systems in place to reduce the effects of a needlestick injury will then be discussed. This will include a review of vaccinations, post exposure prophylaxis and the provision of psychological support – the effectiveness of which will be evaluated.

Definition of terms

It is important to clarify the terminology used within this chapter so that ambiguity may be avoided. The key terms which will now be defined are 'needlestick injuries, 'cause' and 'effect'. Despite an abundance of literature relating to the subject of needlestick injuries, there are very few authoritative definitions available. This is possibly because the term is used so frequently that an assumption is made that everyone knows what it means.

Bandolier (2003) suggests that a needlestick injury is "the introduction of 'blood or other potentially infectious material" into the body of a healthcare worker by a "hollow bore needle or sharp instrument including, but not limited to, needles, lancets, glass and contaminated broken glass". A less detailed but similar definition is provided by the Safer Needles Network (2005), who state that needlestick injuries occur when "healthcare workers jab themselves or a colleague with a needle, or other sharp medical device which is contaminated with potentially infected blood". In contrast, the Canadian Centre for Occupational Health and Safety (2005) simply describes needlestick injuries as "wounds caused by needles that accidentally puncture the skin". An Internet search using www.google.com found the following definition: "penetrating stab wound caused by a needle."

These latter two definitions imply that a needlestick injury could be caused by a clean needle, which is very different from the first two definitions which make it clear that a needlestick injury involves a source with potential infection. This disparity over the meaning of the term 'needlestick injury' highlights the confusion and discrepancy that might occur especially when reporting and recording needlestick injury data.

The terms cause and effect are less controversial to define. Causes are concerned with 'why things happen' whilst effects are 'what happens as a result' (The Writing Centre, 2005). Similarly, Walters (2000) describes a cause as the *reason* for something whilst the effect is the *result* of something happening. Longman (1984) incorporates the two terms within a single definition: 'a cause is that which brings about an effect or a result'

Incidence of needlestick injuries

NHS Employers (2005) acknowledge that needlestick injuries are a significant issue for the NHS, with 400,000 needlestick injuries being reported each year. They add, however, that this figure could reflect significant under reporting since, they say, that there are at least as many incidents that also take place that are not reported.

In May 2000, the Royal College of Nursing (RCN) launched their campaign "Be sharp, Be safe" (Gabriel, 2004). A key activity of the campaign was to determine the extent of needlestick injuries, whilst at the same time evaluating an electronic sharps injury data collection system called EPINet. Fourteen NHS trusts were included in the study, although only 12 submitted their final figures (EPINet, 2005). During the first study (June 2000 to July 2001) there were 888 recorded needlestick injuries, whilst in the second study, which ran from January 2002 to December 2002, there were 1,445 needlestick injuries reported (Gabriel 2004). It is disappointing to note that the information provided by EPINet on their webpage covers only the first six month of the second study, despite the full results being available via the RCN (2003) and NHS Purchasing and Supply Agency (2005) websites.

Data from both studies showed that nurses account for the highest percentage of healthcare workers sustaining needlestick injuries (Gabriel, 2004), although this would be expected, given that they comprise 44% of the NHS workforce and perform many of the clinical procedures involving sharp devices (May, 2002). Table 9.1 shows the job category of the injured worker from the first study, whilst figure 9.1 shows that there is very little difference between the two studies of the groups of workers injured (Gabriel, 2004).

Job Category	Incidence of needlestick injury
Nurses	43%
Junior Doctors	16%
Consultants/ Registrars	8%
Healthcare Assistants (HCA)	7%
Students (med./nursing)	5%
Domestics / Porters	4%
Phlebotomists / IV teams	3%
Sterile Services	2%
Others	11%

Table 9.1 – Job category of the injured workers. (Source: May, 2002)

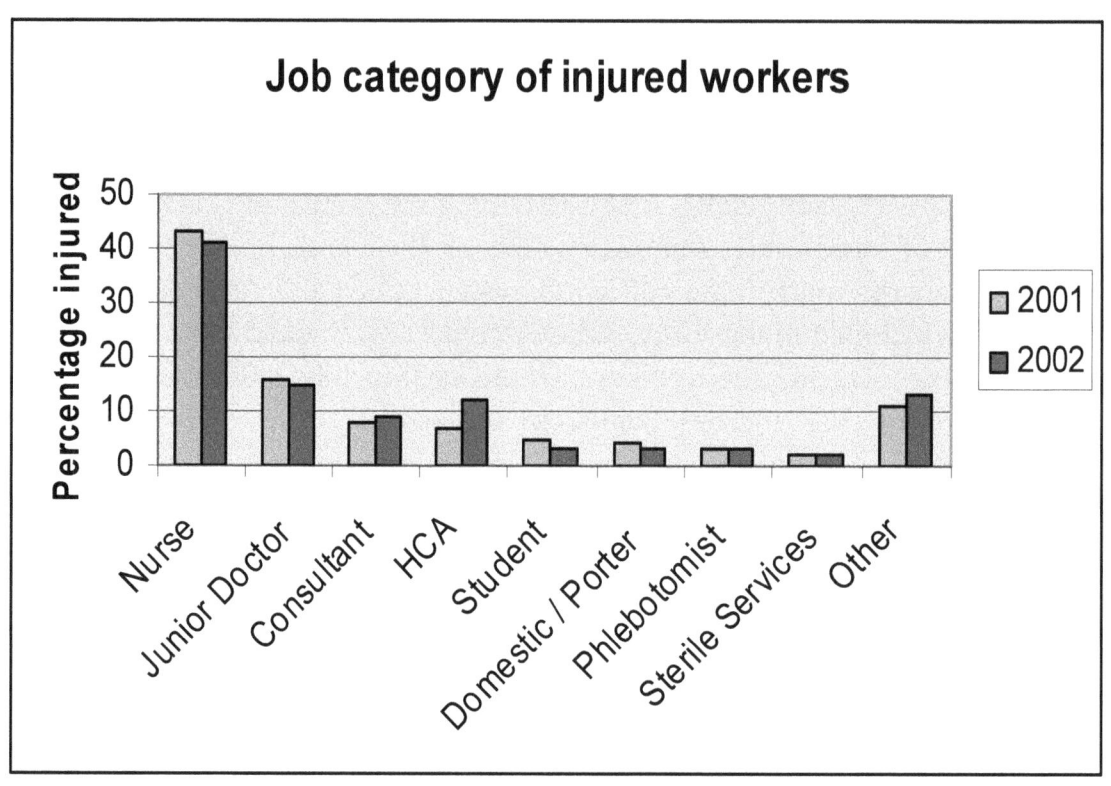

Figure 9.1 – Job category of the injured workers: comparison 2001 and 2002
(Source: RCN, 2003)

May (2002) discusses whether the person sustaining the needlestick injury was the original user of the device (see table 9.2). Interestingly, a significant proportion of injuries (37%) were sustained by a healthcare worker who was not the original user of the device.

Original user	58%
Not original user	37%
Unknown	4%

Table 9.2 – Identification of person who sustained the needlestick injury
(Source: May, 2002)

In September 1999, all Scottish NHS trusts and health boards were asked by the Scottish Executive Health Department to provide information relating to the incidence of needlestick injuries in their area. The numbers of needlestick injuries had increased by 12.5 per cent over the three years that were analysed (1997 – 1999) from 2,168 to 2,439 (Scottish Executive, 2001). However, data were not collected from all NHS Trusts in Scotland and figures from primary care were excluded, so these figures do not provide an accurate picture of needlestick injuries in Scotland. Nonetheless, needlestick injuries were highlighted in the strategy document for NHS Scotland as one of the most common types of injury to staff in NHS Scotland. Consequently, a needlestick injury working group was established in March 2000 to investigate the prevalence, cause and prevention of such injuries and to make recommendations to minimise the risk to staff .

In the USA, the exact numbers of incidents are unknown, but it has been estimated that between 600,000 and 800,000 needlestick injuries occur each year (Bandolier 2003). It has also been estimated that over a 20 year career, 60 per cent of nurses will have at least one needlestick injury (Clarke et al, 2002). In 2000, the needlestick safety and prevention act was signed into law in an attempt to protect healthcare workers from accidental needlestick injuries, by requiring healthcare facilities to review and make available safety-engineered sharps products (Jenkins, 2000).

Based on reported incidents, the International Council of Nurses (ICN, 2000), which is based in Switzerland, believes American health workers suffer 800,000 to 1 million needlestick injuries annually, but this could be a significantly less that the real figure due to under reporting (The ICN also estimates that there are more than 100,000 needlestick injuries in UK hospitals each year. Wilburn (2004), however, provides a more positive picture for the USA, claiming that needlestick injuries have decreased to 385,000 incidents in the year 2000, but states that this is still a significant and disturbing statistic.

Needlestick injury is also an issue in wider Europe. In France, a survey was carried out between 1995 and 1999 among 28 hospitals in Paris, with over 62,000 employees. Approximately 10,563 blood exposures were analysed and 75 per cent of the incidents were due to needlestick injuries (Gehanno, 2001). In Italy, over a five and a half year period, there were 19,860 needlestick injuries reported from a total of 41 hospitals and, once again, nurses were the staff group most likely to receive a needlestick injury, being exposed in 57 per cent of incidents (Ippolito et al, 1999). And in Belgium, and one study based in 22 hospitals found that 33 per cent of healthcare workers had received at least one needlestick injury (Moens et al, 2000).

Causes of needlestick injuries

To facilitate an understanding of the causes of needlestick injuries, it is useful to determine what procedure is most often being undertaken when a needlestick injury occurs. It is also helpful to ascertain exactly when, during this procedure, the injury occurs and which devices are commonly responsible for the injuries sustained. This information may be particularly valuable for the development of prevention strategies and is summarised below.

Table 9.3 identifies procedures being undertaken at the time of a needlestick injury and highlights that healthcare workers undertaking injections or venepuncture with hollow bore needles are at a higher risk of needlestick injury than staff carrying out other procedures.

Procedure	Incidence of injury
Injection (IM/SC)	22%
Vene/arterial puncture	17%
Suturing	10%
Cutting	6%
Cannulation	5%
Finger / heel prick	4%
Subcutaneous infusion	3%
Other needle activity	6.5%

Table 9.3 – Procedures undertaken when a needlestick injury is sustained. (Source: May, 2002)

The USA national surveillance system for healthcare workers has identified six devices that are responsible for the majority (nearly 80 per cent) of needlestick injuries sustained and these are shown in figure 9.2.

The devices involved in needlestick injuries which are of particular concern are hollow-bore needles, especially those used for blood collection or intra venous catheter insertion. These devices are likely to contain residual blood and are associated with an increased risk of transmission of blood-borne pathogens (Cardo et al, 1997). Unfortunately, as can be seen from figure 9.2, the most common device involved in needlestick injuries are hollow bore needles (59 per cent). These findings replicate those of the Health Protection Agency (2005a), who discovered that the majority of needlestick injuries to health care workers in England involve hollow bore needles (63 per cent).

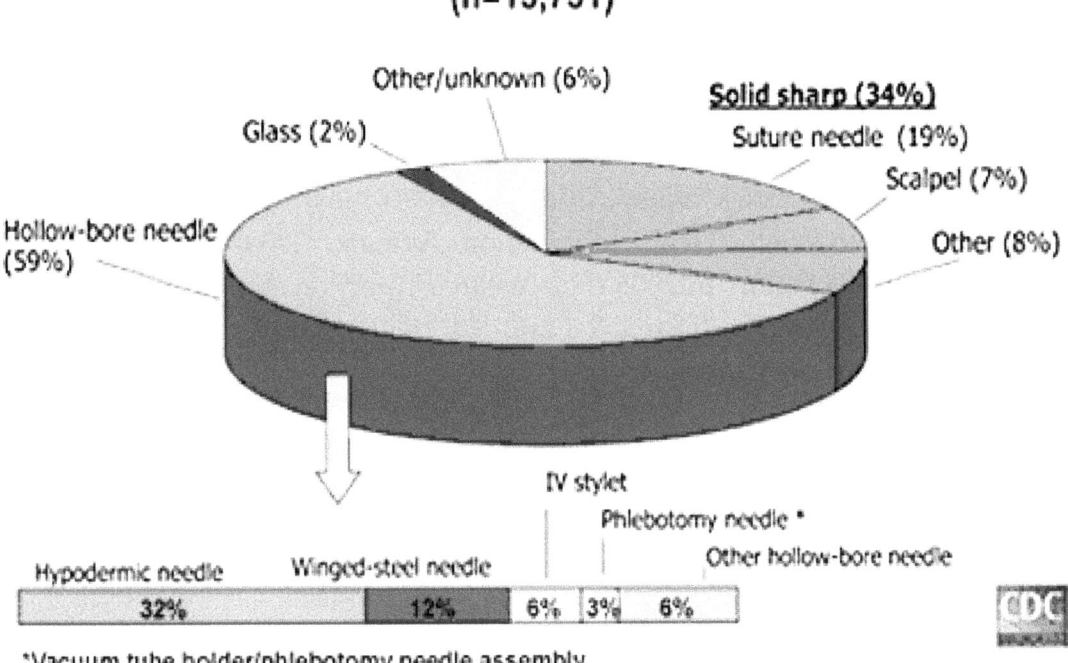

Figure 9.2 - Devices involved in needlestick injuries
(Source: Centres for Disease Control and Prevention, 2004)

Table 9.4 highlights when needlestick injuries occur; and it can be seen that the highest risk of injury occurs during the procedure and before disposal of the sharp.

Activity	Incidence of injury
During use	36%
After use, before disposal	20%
Placing in / protruding from container	10%
Device left in inappropriate place	7%
Disassembling device	5%
Between steps in procedure	3%
Recapping	3%
Protruding from waste bag	3%
Preparation for re-use	2%

Table 9.4 – Activity undertaken when a needlestick injury occurs.
(Source: May, 2002)

It is also useful to determine the location where needlestick injuries take place, since this may assist the prioritisation of resource allocation to reduce these injuries. The Health Protection Agency (2005a) found the majority of needlestick incidents occurred in the ward area, with 45 per cent of incidents reported, followed by operating theatres (15 per cent), accident and emergency departments (11 per cent) and intensive care units (eight per cent). Together, these locations account for 78 per cent of all needlestick injuries reported.

The Health Protection Agency have also analysed the contributory factors attributed to a needlestick injury. In 2004, the 10 most common contributory factors were published (Health Protection Agency, 2004) and these were rationalised to five factors in 2005 (Healthcare Protection Agency 2005a). These results are summarised in figures 9.3 and 9.4. (Note that in these figures, 'UP' refers to 'universal precautions').

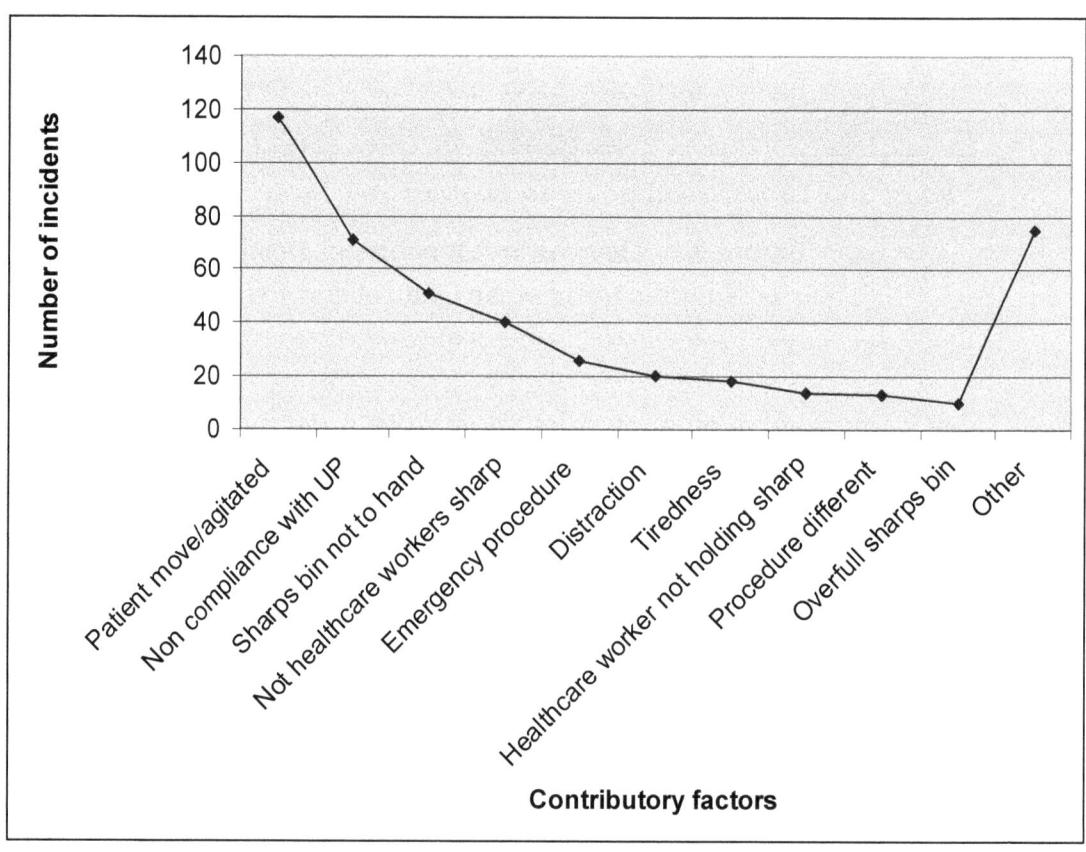

Figure 9.3 – Ten most common contributory factors to a needlestick injury
(Source: Health Protection Agency, 2004)

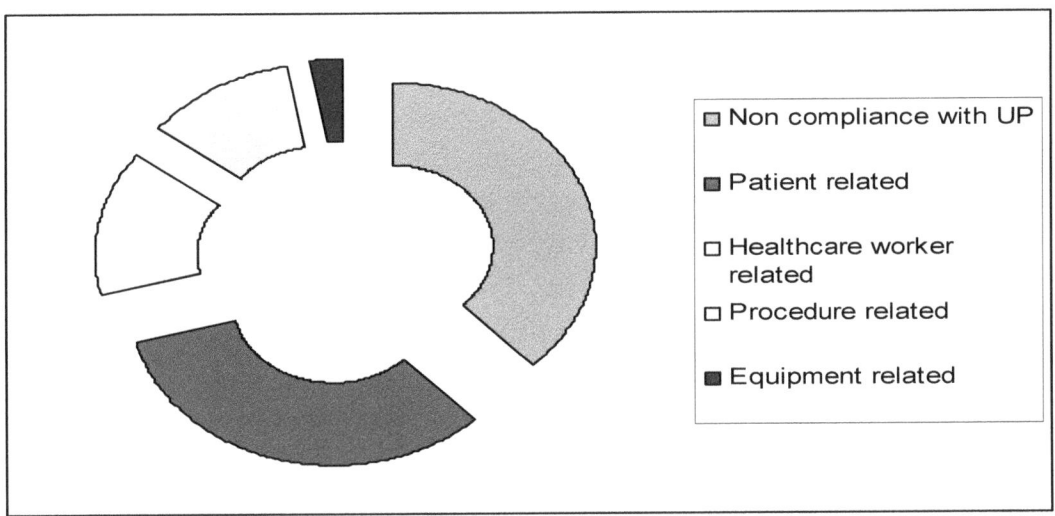

Figure 9.4 – Five most common contributory factors to a needlestick injury
(Source: Health Protection Agency 2005a)

It can be seen that those factors most likely to have a significant impact on the incidence of a needlestick injury are those relating to non compliance with universal precautions (UPs), factors relating to patient activity during a procedure and the attributes of the healthcare worker throughout the process. Clarke et al (2002) also found that high workloads were associated with 50 per cent to two-fold increases in the likelihood of needlestick injuries. This knowledge may help account for the contributory factors relating to the healthcare worker that have an impact when they are distracted, tired and have limited time to prepare for and perform a procedure using a device likely to lead to a needlestick injury.

A review of systems used to reduce the causes of needlestick injuries

It is believed that many needlestick injuries are preventable through adherence to universal precautions (Health Protection Agency, 2005a). Universal precautions are described as a set of guidelines that aim to protect health care workers from blood-borne infections (Bennett and Mansell, 2004); and they should be applied to every patient receiving medical care, regardless of their presumed infection status (Health Protection Agency, 2005a). These guidelines include the use of personal protective equipment (PPE) to prevent blood and body fluids from reaching the worker's skin, mucous membranes, or personal clothing. This may consist of gloves, lab coats, gowns, aprons, shoe covers, goggles, glasses with side shields and masks (Canadian Centre for Occupational Health and Safety, 2002). Although PPE cannot eradicate risk, it can reduce the amount of blood introduced into a puncture wound by a wiping effect as the needle goes through the protection. In particular, current guidance recommends that gloves should be worn whenever blood contact is anticipated (May, 2002).

Universal precautions also include the use of work place practice controls. This requires the development of practical techniques that reduce the likelihood of exposure by changing the way a task is performed (Canadian Centre for Occupational Health and Safety, 2002). Examples of such activities include washing hands before and after procedures, changing gloves between patients, covering any skin lesions with waterproof dressings and the safe handling and disposal of sharps. Dealing safely with sharps includes using readily accessible puncture-resistant sharp containers that are closed, sealed and destroyed before they are completely full, facilitating disposal of sharps immediately after use and not leaving clinical waste for others to clear (World Health Organisation, 2005; Health Protection Agency, 2005a).

The Health Protection Agency (2005b) provides posters which highlight good and bad practice which should be followed to reduce the incidence of needlestick injuries. They focus specifically on the correct disposal of sharps into a sharps bin and the absolute requirement never to recap needles. However, as May (2002) warns, when reflecting upon the large proportion of injuries sustained by healthcare workers who were not the original users of the sharp device, that by discouraging the practice of re-sheathing to protect the user of the device, the injury risk may have been inadvertently transferred to others.

Universal precautions are not always incorporated into daily work practice. For example, Trim et al (2003) evaluated the knowledge of 200 health care workers relating to the implementation of universal precautions. They discovered that, despite a comprehensive educational programme for nurses and training for medical staff, polices and procedures were not followed and gloves were not routinely worn in the clinical setting.

Similarly, Bennett and Mansell (2004) explored community nurses' experience and practice of using universal precautions. They found that, whilst the majority of respondents reported compliance, a small number stated that they re-sheathed needles, inappropriately stored sharps containers and inadequately wore gloves.

In November 2000, a new era for needlestick prevention began when President Bill Clinton signed the needlestick safety and prevention act shifting the focus in exposure from behaviour to devices (Wilburn 2004). The law became effective in April 2001 and made explicit the requirement for health care settings in the USA to identify, evaluate, and make use of effective safer medical devices (Congress, 2000).

One such device is the safety syringe, which is a single use syringe with a retractable needle which goes into the plunger following use (Mahurker, 2005). However, there are many other types of safety medical devices available that include protected needle intravenous connectors, hinged or sliding shields attached to phlebotomy needles, protective encasements to receive an IV stylet as it is withdrawn from the cannula, and safer IV cannulas that encase the needle after use (Occupational Safety and Health Administration, 1999).

Clarke et al (2002) found that the use of any type of protective equipment for taking blood and administering intravenous infusions was associated with a 20 to 30 per cent decrease in needlestick injuries, whilst Mendelson et al (2003) discovered that the implementation of safety needles substantially reduced needlestick injuries amongst healthcare workers. Similarly, Marini et al (2004) reported a marked decrease in exposures to blood-borne pathogens following the implementation of a needle-less system and intravenous safety devices in a children's hospital in Boston, USA. However, they also reported that needlestick injuries do still occur; and this is primarily as a result of a lack of familiarity with the device so that the safety mechanism is not activated, and of staff resistance due to their lack of expertise with the device. The demand for safety devices in the UK has been amplified by the needlestick safety and prevention act in the USA (Frost and Sullivan, 2004).

A search on the Internet search engine www.google.co.uk for "safety syringes" in December 2005 produced over 40,000 'hits'. Many of the available sites are from manufacturers endeavouring to promote their products. Despite technological innovations, however, the market for safety syringes in Europe has been limited. This is partly due to a lack of legislation pertaining to regulations for employee safety, and partly due to the high price of these devices (Frost and Sullivan, 2004). However, there have been some attempts to introduce safety devices into the UK.

The National Audit Office (2003) says that 14 per cent of acute and ambulance trusts are trialling the use of alternative safer needles, and describe the introduction of a safer needle system at a London NHS Trust. Potential costs of time staff spent away from work, locum staff, occupational health input and treatment were calculated, and it was felt there was the potential to make £55,000 savings if the incidence of needlestick injuries could be reduced (see table 9.5). Although the purchase of safer devices was estimated to cost around £136,000 (£81,000 more than the anticipated saving) the trust considered the unquantified impact of needlestick injuries in terms of fines for breaches of health and safety legislation and potential litigation costs. Unfortunately, there is no follow up report looking at the success of this venture.

ITEM	COST	ITEM	BENEFIT
Purchase of retractable cannulas	£136,000	Reduced needlestick injuries (saving staff costs for sick pay, replacement staff, occupational health input and treatment).	£55,000
		Saving of potential litigation costs	£500,000
TOTAL	**£136,000**	**TOTAL**	**£555,000**

Table 9.5 - Cost benefit analysis for the introduction of a safer needle system (Source: National Audit Office, 2003)

Some evidence from the UK about the effectiveness of safety devices does exist, albeit at the present time the evidence is limited. For example, Zakrsewska et al (2001) introduced safety syringes into a UK dental school and discovered that the new device was instrumental in reducing their average of needlestick injury rate from 11.8 to zero per 1,000,000 hours worked per year over a trial period of two years.

However, the United States General Accounting Office (2000) offers a word of warning about viewing the use of safety devices as a panacea, stating that these devices do have limitations. For example, those that have been assessed vary considerably in their clinical efficacy and effectiveness in reducing rates of injuries. In some cases, these devices have even caused needlestick injuries. Besides these limitations, there are other obstacles to the use of needles with safety features, which include possible staff resistance to changes in the devices used, and the time required to train staff in the use of new devices (United States General Accounting Office, 2000).

Table 9.4 shows that 36 per cent of injuries occur "during use" and this may occur as a result of inexperience, poor technique or lack of training. Similarly, injuries occurring as a result of poor practice (13 per cent), either between steps in a procedure, disassembling, whilst preparing for re-use, or re-sheathing, suggest that training and education of practitioners' in clinical skills requires a fundamental review (May, 2002).

The World Health Organisation (2005) also supports training for health care workers, highlighting the need for staff to be educated about occupational risks and the need to use universal precautions. In addition, Eucomed (2001) believe that conducting regular refresher training is crucial, as complacency can become a contributing factor with experienced healthcare workers, who can develop the attitude that such risks are "normal" and "expected" for their profession.

Reddy and Emery (2001) evaluated the effects on the incidence of needlestick injuries of an extensive educational programme. The programme commenced in 1995 and informed all hospital employees of the importance of needlestick safety and blood-borne pathogens. In 1997, the use of safety syringes and needle less intravenous systems was implemented in all departments. Over the six years reviewed, the incidence of needlestick injuries per 100 full time employees fell from ten per cent to four per cent (Figure 9.5).

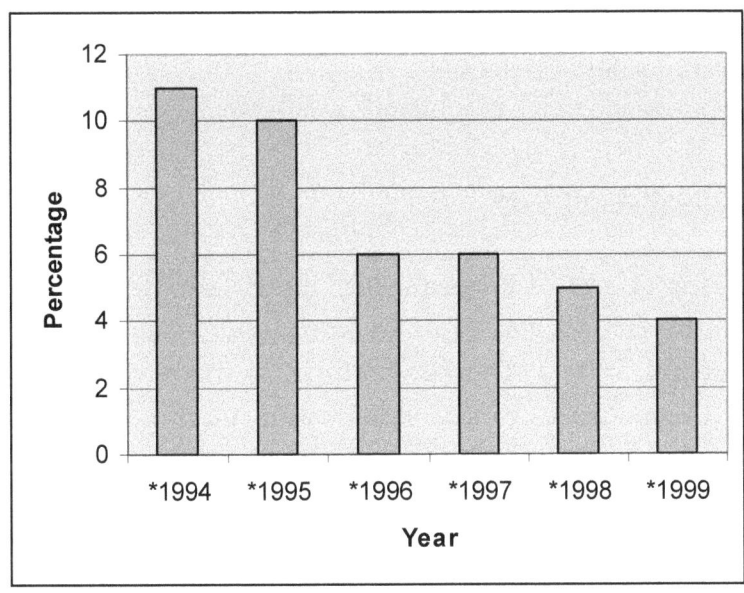

Figure 9.5 – Incidence of needlestick injuries per 100 full time employees
(Source: Reddy and Emery, 2001)

A further strategy to reduce the incidence of needlestick injuries is to increase awareness of injuries that are happening both in the health service and within the employee's local organisation. This approach will help to bring home the need to be cautious when dealing with devices that can lead to an injury. This approach may also encourage needlestick incidents to be reported since under-reporting of incidents is commonplace. This is highlighted by Elmiyeh et al (2004), who discovered that, although 80 per cent of respondents were aware that needlestick incidents should be notified, only 51 per cent of those affected had reported all needle-stick injuries.

The World Health Organisation (WHO) also recommends reducing nonessential procedures so that health care workers are trained to avoid unnecessary blood transfusions (e.g. using volume replacement solutions), injections (e.g. prescribing oral equivalents), suturing (e.g. reduction of episiotomies) and other invasive procedures (WHO, 2005).

The United States General Accounting Office (2000) estimates that approximately 29 per cent of needlestick injuries in hospitals can be prevented each year through the use of needles with safety features. They also calculate that a further 46 per cent can be prevented by eliminating the use of unnecessary needles, education and safer working practices. This view is supported by evidence from Jagger (1996) who found a 59 per cent reduction in needlestick injuries following an education programme and implementation of universal precautions. Subsequent to a period of consolidation, a further reduction of 84 per cent in injuries was discovered following the implementation of safety devices.

Effects of needlestick injuries

The most serious effects of a needlestick injury are Hepatitis B Virus (HBV), Hepatitis C Virus (HCV) and Human Immunodeficiency Virus (HIV) (Royal College of Nursing, 2005a). The key features of these viruses are explained in appendix 9.1. These viruses can be serious and life threatening and the risks of infection following a needlestick injury with a contaminated needle have been estimated as and are tabulated in table 9.6.

Occupational Exposure	Risk of Transmission
HIV	0.3 per cent (1 in 300 chance of infection)
Hepatitis B Virus (HBV)	2 – 40 per cent
Hepatitis C Virus (HCV)	2.7 – 10 per cent

Table 9.6 – Risk of infection following a needlestick injury with contaminated needle (Source: Wilburn, 2004)

As at December 2001, there were 57 documented cases and 138 possible cases of occupationally acquired HIV infection among healthcare personnel in the United States since reporting began in 1985 (Centres for Disease Control and Prevention, 2003). Bandolier (2003) reports that needlestick injuries result in at least 1,000 new cases of health care workers being diagnosed with HIV, HBV or HCV every year in the USA.

The first documented HIV seroconversion due to an occupational exposure in a healthcare worker occurred in the UK in 1984 and at least four UK healthcare workers are known to have died following occupationally acquired HIV infection (NHS Employers 2005). There were three reports of seroconversions for HCV in the five years since the start of enhanced surveillance in 1998. However, in the past 12 months, another six seroconversions have been reported to the Health Protection Agency (2005a). Most of these involved percutaneous injuries of moderate depth from hollow bore needles contaminated with fresh blood from source patients who were mostly intravenous drug users. Almost two-thirds of the seroconversions were preventable since they resulted from injuries caused by non-compliance with universal precautions (Health Protection Agency, 2005a). These figures are regarded as an underestimate due to under-reporting of needlestick injuries, since HCV, like HIV, may not be recognised unless the healthcare worker reports the injury and has the relevant blood test (NHS Employers, 2005).

In addition to the risk of acquiring a seriously debilitating or fatal disease, the psychological impact of a needlestick or other sharps injury can be very significant. A lengthy process of diagnostic procedures must be followed before it is known whether a serious disease has been contracted or not (Eucomed, 2001). Not knowing the infection status of the source patient can accentuate the healthcare worker's stress (Bandolier, 2003). In one study of 20 health care workers with an HIV exposure, 11 reported acute severe distress, seven had persistent moderate distress, and six left their jobs as a result of their exposure (International Council of Nurses, 2000).

Whilst it is, according to Unison (2002) "impossible to put a cost on the misery caused by infection though needlestick injuries or the agony of health workers and their families waiting to know if an injury will lead to disease", nevertheless compensation claims from individuals injured by needles could run into the tens of thousands of pounds per case.

In 2002, a healthcare worker received an award of £58,000 for a needlestick injury sustained in 1997 by a HBV contaminated needle. Payment was made in respect of the severe shock and trauma suffered by the healthcare worker, who developed needle phobia and was unable to continue work in his previous role.

Interestingly, this worker did not develop Hepatitis B (National Audit Office, 2003).

There are other financial costs linked to the effects of sustaining a needlestick injury including the direct costs associated with the initial and follow-up treatment of exposed healthcare personnel, which are estimated to range from $500 to $3,000 depending on the treatment provided (United States General Accounting Office 2000). There are also longer term costs relating to tests and treatment as well as the loss of valuable trained staff from the workforce through illness (Unison 2002).

An interesting yet alarming study was undertaken by Leliopoulou et al (1999), who investigated nurses' perceptions of risk of contracting infection following a needlestick injury. Forty nine per cent of nurses working in high risk areas such as intensive care, haematology and haemodialysis believed a needlestick injury with contaminated blood was an unlikely sauce of infection. Of equal concern was the finding that 67 per cent of respondents felt that nurses were not at a higher risk of exposure to HIV and Hepatitis B than other health care workers. These findings would suggest that nurses would benefit from further education regarding infection from blood borne viruses.

A review of systems used to reduce the effects of needlestick injuries

According to NHS Employers (2005), testing is recommended for HBV, HCV and HIV following any of the following:

- a healthcare worker to patient blood-exposure incident
- a needlestick injury
- any exposure to body fluids

This is crucial since it is now known that seroconversions in healthcare workers **do** occur.

It is also essential that the importance of urgent reporting of any needlestick injury continues to be stressed. This allows a suitable risk assessment of the incident to be undertaken which, together with obtaining consent for testing of the source patient, will enable appropriate management of the injured healthcare worker. Unreported needlestick injuries are a serious problem since they can prevent healthcare workers from receiving appropriate treatment (Wilburn, 2004).

The risk of contracting HBV from needlestick exposure in a health care setting is much higher than HIV, because the virus is both more infectious and has greater prevalence (Royal College of Nursing, 2005b). As a result, it has been recommended that all nurses be vaccinated against HBV, with a booster being administered when a needlestick injury occurs and when titre levels are low (Royal College of Nursing, 2005b). This recommendation is supported in guidance issued by NHS Employers (2005) that advises testing healthcare workers for HBV infection at the pre-employment check and undertaking immunization if they are likely to be working in environments using contaminated needles and sharps.

Since the HBV vaccine became available in 1982, the annual number of occupational infections of Hepatitis B in the USA has decreased 95 per cent from more than 10,000 in 1983 to less than 400 in 2001 (Centers for Disease Control and Prevention, 2003). Despite this information, according to the World Health Organisation, in some areas of the world, over 80 per cent of healthcare workers have not been immunised against HBV (Pruss-Ustun et al, 2003).

Closer to home, Alzahrani et al (2000) undertook a retrospective examination of 2646 needlestick incidents in 10 hospitals in the Greater Manchester area. They found that 10 per cent of the staff injured in these incidents had never been vaccinated against HBV, and 27 per cent of those who had been vaccinated had no detectable antibodies to HBV.

In the case of HCV, there is currently no vaccine or chemoprophylaxis. As a minimum, appropriate testing at correct time intervals is important in facilitating the early detection of HCV infections, and a prompt referral for specialist advice. It has been shown that healthcare workers who have recently seroconverted and are started on treatment within six months of their infection go on to clear the virus and do not progress to chronic HCV (Health Protection Agency, 2005a). Fortunately, eight of the nine healthcare workers with occupationally acquired HCV infection have shown evidence of having cleared the virus, six following early treatment with antiretrovirals (Health Protection Agency, 2005a).

The Department of Health (DH) have produced an action plan to deal with the incidence of HCV and achieve a reduction in morbidity and mortality relating to this infection (Department of Health, 2004). There are four actions contained within the guidance, which address surveillance and research, increasing awareness and reducing undiagnosed infections, providing high quality health and social care services and prevention. There has yet to be an evaluation undertaken to show the effectiveness of these action areas.

For those healthcare workers who experience significant occupational exposure to HIV source patients, it is important that they commence HIV post exposure prophylaxis promptly (Health Protection Agency, 2005a). It was discovered that 77 per cent of healthcare workers commenced prophylaxis within 24 hours of exposure, and that 38 per cent of these were within one hour, consistent with national guidance (Health Protection Agency, 2005a). When it was found that the exposure had been to an HIV negative source patient, 90 per cent of the healthcare workers had been taking prophylaxis for a week or less, with 54 per cent of these only on the prophylaxis for one day. However, some healthcare workers were still taking toxic drugs with unpleasant side effects inappropriately because of apparent delays in source patient HIV testing (Health Protection Agency, 2005a). Wilburn and Eijkemans (2004) report that post exposure prophylaxis with antiretroviral medications can reduce the risk of HIV transmission by 80 per cent.

To further ensure the effects of needlestick injuries are minimized, it is essential that all healthcare workers are aware of local post-exposure prophylaxis policies and procedures, in particular the need for prompt action following a known or potential exposure to HIV (NHS Employers, 2005).

To help deal with the psychological effects of needlestick injuries, a free phone hotline service has been established in New South Wales, Australia. It is available 24 hours a day, seven days a week. Its intention is to provide an information, support and referral service for healthcare workers who sustain a needlestick injury. The hotline supplements the local management of needlestick injuries by providing risk assessments, prophylaxis information and counselling support (South East Sydney Area Health Service, 2005). There is currently no research available to evaluate this service but it can be anticipated that the principle will be valued. There is no evidence of such a service in the UK, where the psychological support following a needlestick injury appears to be limited, if not overlooked, and receives limited mention in the literature.

Conclusion

As this chapter has shown, needlestick injuries continue to be a serious hazard, exposing health care workers to deadly viruses and other blood borne pathogens despite significant progress in policy, practice and products (Wilburn, 2004). Greater collaborative efforts by all stakeholders are needed to prevent needlestick injuries and the tragic consequences that can result. Such efforts are best accomplished through a comprehensive program that addresses institutional, behavioural, and device-related factors that contribute to the occurrence of needlestick injuries in health care workers (National Institute for Occupational Safety and Health, 1999).

Critical to this effort is the continuing commitment to:

- Eliminate needle bearing devices where safe and effective alternatives are available and the development, evaluation, and use of needle devices with safety features (National Institute for Occupational Safety and Health, 1999).
- Ensure proper adherence to universal precautions and adequate education and training. Healthcare workers should know the current protocols and guidance on what to do to avoid exposures and manage them when they occur (Health Protection Agency, 2005a).
- Review current work practices and pursue the elimination of unnecessary injections and unnecessary sharps (Wilburn, 2004).
- Encourage reporting of all needlestick injuries so that, with knowledge of the circumstances of each incident and the device responsible, positive action can be taken to minimise the recurrence of injuries in the future

It is also important to highlight the responsibility that nurses have to themselves, as well as to patients and society as a whole, to minimize the potential for the spread of blood-borne infections by adopting safer working practices and using safety technology (Gabriel, 2004).

Finally, although immunisation and prophylaxis can be used to reduce some of the effects of needlestick injuries, it is critical to raise awareness that these interventions are no substitute for effective needlestick injury prevention working practices (Health Protection Agency, 2005a).

Appendix 9.1 – Key Features Of HBV, HCV And HIV

(Sources: American Nurses Association, 1999; Bandolier, 2003; and Department of Health, 2004)

Hepatitis B (HBV)

HBV has infected approximately two billion people in the world, roughly a quarter to a third of the world's population. About 300 million people are carriers of the virus. Whilst the carrier rate is low in most western countries (less than one per cent in the UK and USA, for example), in Africa and some parts of Asia the carrier rate can be well above 10 per cent.

Spread of HBV is often by intravenous routes through infected blood or blood products, or contaminated needles used by drug abusers, tattooists or acupuncturists. Another major route is close personal contact, with the virus being present in semen and saliva. Perhaps the most important transmission route worldwide is vertical transmission from mother to baby.

While most infected persons recover completely, Hepatitis can occur in up to one per cent of cases, and some patients go on to develop chronic hepatitis or liver cancer. Some people become carriers, which may preclude them from working in their chosen career if, for example, they are healthcare professionals.

Regulatory and legislative efforts, including the Occupational Safety and Health Administration (OSHA) Blood Borne Pathogens Standard, have largely been responsible for the reduction of deaths from Hepatitis B as a result of vaccine programs. In addition, cases of Hepatitis B in health care workers have dropped from 17,000 to 400 annually–and continue to drop.

Vaccination against HBV should normally be given to healthcare personnel in the UK, including members of emergency and rescue teams, people with haemophilia, and some other higher risk conditions or professions.

If a healthcare worker is exposed to HBV and receives post exposure treatment, it is unlikely that the person will become infected and pass the infection on to others.

Hepatitis C (HCV)

HCV was only identified in 1988, and has since been found to be responsible for the majority of post-transfusion hepatitis.

The rate of infection is about 0.02 per cent in northern Europe, but is six per cent in Africa and as high as 19 per cent in Egypt. Incidence is high in intravenous drug users and people with haemophilia because transmission is through blood or blood products, and through vertical transmission from mother to child.

Infection is mostly asymptomatic, with one in 109 infected people having an influenza-like illness with jaundice. Most patients are detected when they present many years later with chronic liver disease, which occurs in about half of infected patients, with cirrhosis and liver cancer being common. Determining the actual prevalence of HCV infection and liver disease depends upon good epidemiology. There could be thousands of nurses with occupationally acquired hepatitis C who do not know it.

In cases of needlestick injury, antiviral agents may be given, but there is no good evidence that they prevent infection.

HCV infection in a healthcare worker may result in loss of employment because of the risk of transmission of HCV to uninfected patients. These risks have been estimated at 50 per cent likelihood of one patient being infected in 5,000 procedures carried out by an HCV infected surgeon over 10 years.

HCV is eight to 10 times more common than HIV and is five to 10 times more transmissible than HIV from needlestick injuries.

An estimated 0.5 per cent of the general population in England (approximately 250,000 people) have been infected with HCV. About 20 per cent of those infected appear to get rid of the virus naturally without treatment. Thus, 0.4 per cent of the population (some 200,000 people) are chronically infected with Hepatitis C.

However, there have been only 38,000 diagnoses of hepatitis C infection reported, so it is concluded that the majority of infected people are undiagnosed.

Moderate to severe disease can now be treated successfully in up to 55 per cent of cases overall. If chronic infection is left untreated, most people will eventually develop symptoms, and one in five will go on to develop cirrhosis of the liver after 20 years or more. A small number will develop primary liver cancer.

Human Immunodeficiency Virus (HIV)

There are about 35 million known cases in the world, but it is commonly recognised that this is likely to be a gross underestimate. Many cases are in Africa, where prevalence can be very high. Other areas with high prevalence include parts of Asia, and parts of western countries where needle-sharing or sexual practices increase the risk. Vertical transmission from mother to baby is high, unless specific treatments are instituted.

Most HIV seroconversions are clinically silent, though some might be associated with a short self-limiting illness. After a symptom free period which is often many years in otherwise healthy individuals, symptomatic HIV infection is associated with increasing viral load and failure of the immune system. HIV is the virus that causes AIDS, a fatal disease. Advances in treatment prolong the time before HIV becomes AIDS. The drug treatment can cost up to $6,000 per month.

After needlestick exposure from known (or suspected) HIV infected material, antiviral agents are now commonly used. Treatment with two or even three antiviral agents is likely, though practice varies in different establishments.
During the follow-up period, especially the first 6-12 weeks, the individual should follow the recommendations for preventing the transmission of HIV. These include not donating blood, semen, or organs and not having unprotected sexual intercourse. In addition, women should consider not breast-feeding infants to prevent the possibility of exposing their infants to HIV that may be in breast milk.

References

Alzahrani A, Vallely P and Klapper P (2000). Needlestick injuries and hepatitis B virus vaccination in health care workers. *Communicable Disease and Public Health* 3: 217 – 218.
www.hpa.org.uk/cdph/issues/CDPHVol3/no3/shorts.pdf

American Nurses Association (1999). *Needlestick injury.*
www.nursingworld.org/readroom/fsneedle.htm

Bandolier (2003). Needlestick injuries. *Bandolier Extra.* July, 1-18.
www.jr2.ox.ac.uk/bandolier/booth/booths/needle.html

Bennett G and Mansell I (2004). Universal precautions: a survey of community nurses experience and practice. *Journal of Clinical Nursing* 13 (4) 413 – 421.

Canadian Centre for Occupational Health and Safety (2002). Universal precautions.
www.ccohs.ca/oshanswers/prevention/ppe/universa.html

Canadian Centre for Occupational Health and Safety (2005). What are needlestick injuries?
www.ccohs.ca/oshanswers/diseases/needlestick_injuries.html

Cardo D, Culver D, Ciesielski C, Srivastava P, Marcus R, Abiteboul D, Heptonstall J, Ippolito G, Lot F, McKibben P, and Bell D (1997). A case-control study of HIV seroconversion in health care workers after percutaneous exposure. *The New England Journal of Medicine* 337 (21) 1485-1490.

Centres for Disease Control and Prevention (2003). Exposure to blood – what healthcare personnel need to know.
www.cdc.gov/ncidod/dhqp/pdf/bbp/Exp_to_Blood.pdf

Centres for Disease Control and Prevention (2004). Workbook for designing, implementing and evaluating a sharps injury prevention programme.
www.cdc.gov/sharpssafety/

Clarke S P, Reckett J L and Sloane D M (2002). Organizational climate, staffing, and safety equipment as predictors of needlestick injuries in hospital nurses. *American Journal of Infection Control* 30 (4) 207-216.

Congress (2000). *Needlestick Safety and Prevention Act.* USA.

http://frwebgate.access.gpo.gov/cgi-bin/getdoc.cgi?dbname=106_cong_public_laws&docid=f:publ430.106

Department of Health (2004). *Hepatitis C – Action plan for England*. Department of Health Publications, London.

Elmiyeh B, Whitaker I, James M, Chahal C, Galea A, and Alshafi K (2004). Needle-stick injuries in the National Health Service: a culture of silence. *Journal of the Royal Society of Medicine* 97 (7) 326-327.

EPINet (2005). What is EPINet doing at the moment? www.needlestickforum.net/3epinet/latestresults.htm

Eucomed (2001). Eucomed position paper - Preventing 'sharps' injuries. www.eucomed.be/upload/pdf/tl/position_papers/sharps_injuries_position_paper.pdf

Frost and Sullivan (2004). *Strategic analysis of the European safety syringes market*. www.marketresearch.com/product/display.asp?productid=1058732&g=1

Gabriel J (2004). Reducing the risks of sharps injuries in health professionals. *Nursing Times* 100 (2004) no. 26, p. 28-29.

Gehanno J F (2001). Health care workers and other workers at occupational risk of hepatitis B - a perspective from France. *Viral Hepatitis* October 10 (1) 4-5.

Health Protection Agency (2004). *Surveillance of significant occupational exposure to blood borne viruses in healthcare workers.* www.hpa.org.uk/infections/topics_az/bbv/pdf/6year_analysis.pdf

Health Protection Agency. (2005a). *Eye of the Needle – Surveillance of significant occupational exposure to blood borne viruses in healthcare workers.* www.hpa.org.uk/infections/topics_az/bbv/pdf/eye_of_the_needle.pdf

Health Protection Agency (2005b). *Examples of good and bad practice to avoid sharps injuries.* www.hpa.org.uk/infections/topics_az/bbv/good_bad.htm

Ippolito G, Puro V, Petrosillo N, De Carli G and the Studio Italiano Rischio Occupazionale da HIV (SIROH) group (1999). Surveillance of occupational exposure to blood borne pathogens in health care workers: the Italian national programme. *Eurosurveillance* March, 4 (3) 33-36.

International Council of Nurses (2000). *ICN on preventing needlestick injuries.* www.icn.ch/matters_needles.htm

Jagger J (1996). Reducing occupational exposure to blood borne pathogens: where do we stand a decade later? *Infection Control Hospital Epidemiology* 17 (9), 573-575.

Jenkins C (2000). *BD Celebrates Healthcare Worker Safety Law.* www.bd.com/contentmanager/b_article.asp?Item_ID=21328&ContentType_ID=1&BusinessCode=20001&d=us&s=press&dTitle=Press&dc=&dcTitle=

Leliopoulou C, Waterman H, and Chakrabarty S (1999). Nurses failure to appreciate the risks of infection due to needlestick accidents – a hospital based survey. *Journal of Hospital Infection* 42 (1) 53-59.

Longman (1984). Longman Dictionary of the English Language. Longman Group UK Limited, England.

Mahurker S (2005). *The smart syringe.* www.smartsyringe.com/

Marini M, Giangregorio M, and Kraskinski J (2004). Complying with the occupational safety and health administration's blood borne pathogen standard: implementing needle less systems and intravenous safety devices. *Paediatric Emergency Care* 20 (3) 209 – 214.

May D (2002). *EPINet sharps injury surveillance pilot project. 1 year results July 2000 – June 2001. Preliminary report.* www.pasa.nhs.uk/medicalconsumables/sharps/EPINET_preliminary_report_Jan_2002.doc

Mendelson M, Lin-Chen B, Solomon R, Bailey E, Kogan G and Goldbold J (2003). Evaluation of a safety re-sheathable winged steel needle for the prevention of percutaneous injuries associated with intravenous access procedures among healthcare workers. *Infection Control and Hospital Epidemiology* 24 (2) 105 – 112.

Moens G, Vranckx R, De-Greef L and Jacques P (2000). Prevalence of Hepatitis C Antibodies in a Large Sample of Belgian Healthcare Workers. *Infection Control and Hospital Epidemiology* 21 (3).

National Audit Office (2003). *A safer place to work – improving the management of health and safety risks to staff in NHS Trusts.* The Stationery Office, London.

NHS Employers (2005). *The management of health, safety and welfare issues for NHS staff.*
www.pasa.doh.gov.uk/medicalconsumables/sharps/blue_book_complete%5B1%5D.pdf

NHS Purchasing and Supply Agency (2005). *Sharps and needlesticks.*
www.pasa.nhs.uk/medicalconsumables/sharps/rcn.stm

National Institute for Occupational Safety and Health (1999). NIOSH Alert. *Preventing needlestick injuries in healthcare settings.*
www.cdc.gov/niosh/pdfs/2000-108.pdf

Occupational Safety and Health Administration (1999). *Record Summary of the Request for Information on Occupational Exposure to Blood borne Pathogens Due to Percutaneous Injury.* www.osha-slc.gov/html/ndlreport052099.html

Pruss-Ustun A, Rapiti E and Hutin Y (2003). *Sharps injuries: global burden of disease from sharps injuries to health-care workers. Geneva, Switzerland.* World Health Organisation.
www.who.int/quantifying_ehimpacts/publications/9241562463/en/

Reddy S and Emery R (2001). Assessing the effect of long term availability of engineering controls on needlestick injuries among health care workers: a 3 year pre implementation and post implementation comparison. *American Journal of Infection Control* 29 425 – 427.

Royal College of Nursing (2003). Results of the sharps injury surveillance project. Year 2 – 2002. www.pasa.nhs.uk/medicalconsumables/sharps/rcn.stm

Royal College of Nursing (2005a). Safe handling and disposal of sharps.
http://www.rcn.org.uk/resources/mrsa/healthcarestaff/infectioncontrol/sharps.php

Royal College of Nursing (2005b). *Good practice in infection prevention and control – guidance for nursing staff.* Royal College of Nursing, London.
www.rcn.org.uk/resources/mrsa/downloads/Wipe_it_out-Good_practice_in_infection_prevention_and_control.pdf

Safer Needles Network (2005). *Welcome.* www.needlestickforum.net

Scottish Executive (2001). *Needlestick injuries: sharpen your awareness.*
www.scotland.gov.uk/library3/health/nisa-02.asp

South East Sydney Area Health Service (2005). Needlestick injury hotline. www.sesahs.nsw.gov.au/albionstcentre/needlestick.asp

Trim J, Adams D and Elliott T (2003). Infection control. Health care workers knowledge of inoculation injuries and glove use. *British Journal of Nursing* 12 (4) 215 – 221.

The Writing Centre (2005). The cause and effect essay. www.delmar.edu/engl/wrtctr/handouts/cause_effect.htm

Unison (2002). *Making a point*. www.unison.org.uk/features/features/0211needlesticks.asp

United States General Accounting Office (2000). *Occupational safety: selected cost and benefit implications of needlestick prevention devices for hospitals.* www.gao.gov/new.items/d0160r.p

Walters F S (2000). Cause and effect paragraphs. http://lrs.ed.uiuc.edu/students/fwalters/cause.html

Wilburn S (2004). Needlestick and sharps injury prevention. *Online journal of issues in nursing* 9 (3) Manuscript 4. www.nursingworld.org/ojin/topic25/tpc25_4.htm

Wilburn S and Eijkemans G (2004). Preventing needlestick injuries among healthcare workers. *International Journal of Occupational and Environmental Health* 10 (4) 451 – 456.

World Health Organisation (2005). *Universal precautions, including injection safety.* www.who.int/hiv/topics/precautions/universal/en/

Zakrsewska J, Greenwood I, and Jackson J (2001). Introducing safety syringes into a UK dental school – a controlled study. *British Dental Journal* 190 (2) 88-92.

This page is intentionally blank.

10

The ageing National Health Service workforce: a significant risk to the NHS and to the nation?

CAROLE MODERATE

Introduction

The aim of this chapter is to review the available literature and evidence that supports the demographic trend of an ageing population and an equally ageing nursing work force, and test the hypothesis that:

"An ageing nurse workforce represents a significant risk to the NHS and the nation"

Potential areas of risk discussed are:

- Recruitment and retention
- Physical ability
- Manual handling
- Training and professional development
- Performance, understanding experience in nursing – Novice to expert
- New ways of working / managing change, new technology

Older workers

Older workers are commonly defined as being those over the age of 50 (Ilmarinen, 1997). According to figures from the National Audit Office (NAO, 2004), the general population of the United Kingdom is ageing; in 2004 some 34% (20 million people) were 50 or over. This figure is set to increase to 25.5 million by 2022, with the percentage population over 50 rising to 40%. A fall in the birth rate means fewer young people are entering the work force and improvements in health mean more people are reaching retirement age, it follows therefore, that the working population is also ageing.

In the future, it is likely that more people will need to continue working later in life due to financial constraints. Available pensions, and early retirement packages employers are able to offer, are decreasing in value. People are also more likely still to have financial commitments such as mortgages later on in life. It therefore seems logical that, if there are fewer younger people available in the labour market and an increase in older people who want to continue in work, employers will need to consider how to go about attracting and accommodating an older workforce. The government's Code of Practice for Age Diversity in Employment, launched in June 1999, has focused attention on age issues and encouraged employers to consider developing age-proof policies and practices.

The demographic trend of an ageing workforce is reflected across many international countries. The main two reasons for this are, the baby boom following the Second World War and the low birth rates of the 1980s. It has been predicted, based on current mortality and birth rate tables and emigration rates, that by 2025 the proportion of individuals over the age of 55 years will be 32% in Europe, 30% in North America, 21% in Asia, and 17% in Latin America (Ilmarinen, 1997).

This global demographic trend poses a challenge to all areas of employment including the National Health Service (NHS), and is reflected by the nature of available literature on the subject. Themes emerging from the literature appear to have many commonalities, including the recruitment and retention of older workers and the importance of addressing their particular needs. Therefore, solutions to, and strategies to manage, the risks associated with an ageing workforce in general should be considered in the context of the NHS. Furthermore, there seems to be little research available on the actual risks associated with employing an older nursing workforce; and how organisations are meeting the needs of the older nurses. Much of what is available deals with the demographic challenges of recruitment and retention rather than the risks associated with the older nurses themselves. This is reflected in the literature from America, where the nursing work force is ageing faster than the workforce as a whole, and employers are faced with the same recruitment and retention problems at a time when the international pool of available nurses is declining (Letvak, 2002).

Nurses in the NHS

Nurses are an important resource for the NHS and represent a major investment in terms of what has been spent to acquire and develop them. The organisation's attitude to its employees forms an important context of risk management, as individuals introduce more variability and uncertainty into an organisation than almost anything else (Roberts, 2002).

The register of the Nursing and Midwifery Council (NMC) provides the best source for determining a national overview of the age profile of the nursing and midwifery population in the UK. Statistical trends of the register continue to show an increasingly ageing workforce. Table 10.1 gives a breakdown of the age of registrants over the last decade. In 1995, over half of those on the register were under 40. However, in 2004, more than half of those on the register are over 40; and one in four is over 50. This trend is also reflected in the changing nature of the student nurse with a large proportion being termed as "mature" instead of school leavers.

Table 10.1 - Age distribution of NMC register (%) 1995-2004

	1995	1996	1997	1998	1999	2000	2001	2002	2003	2004
Under 25 yrs	4.45	4.07	3.61	3.17	2.98	2.90	2.12	2.24	2.10	2.02
25-29yrs	14.02	12.79	11.77	10.93	10.32	9.88	7.30	8.86	8.54	8.44
30-39 Yrs	35.92	36.40	36.14	35.71	34.84	33.68	31.31	30.63	29.37	28.30
40-49 Yrs	25.47	26.15	26.66	27.55	28.56	29.58	32.22	32.32	33.26	33.94
50-54 Yrs	9.16	9.43	10.30	10.96	11.28	11.55	12.82	11.46	11.46	11.62
Over 55 Yrs	10.98	11.16	11.50	11.68	12.02	12.31	13.20	14.50	15.27	15.58

A number of factors explain the ageing profile of practitioners on the register. First, the high proportion of nurses in their mid-30s to mid-40s reflects the comparatively large intakes of newly qualified (young) nurses that occurred in the 1970s and early 1980s. Second, intakes of newly qualified practitioners have reduced markedly in recent years. Third, new intakes to nurse education now have a broader age range than was the case in the 1970s and 1980s, when the vast majority of student nurses were school leavers. In addition, a large proportion of nurses who joined the NHS pension scheme before 1995 have retirement rights enabling them to retire with full benefits at 55. Buchan (1998) suggests that by 2010 one in four nurses will be aged 50 or over with one in 10 nurses on the register beyond the trigger age of 55

It would seem, therefore, that one of the main risks of an ageing nursing workforce to the NHS is retaining the nurses we have, and recruiting more. Watson et al (2003) suggests that the NHS has not been devoting enough priority and attention to nurturing older nurses as a valuable resource. This results in older nurses feeling undervalued and not being helped by their employers to address the difficulties associated with growing older in a rapidly changing service. Furthermore, Grey (2003) found that NHS managers' responses to the potential risks of an ageing workforce varied from total ignorance to being seen as a priority. This indicates a national disparity in approaches to this issue, with very few local policies and practices relating specifically to the older workforce.

Ironically, the government's drive for political correctness with equal opportunity policy and practices, in line with the existing voluntary Code of Practice on Age Diversity in Employment, and the proposed age discrimination legislation due to come into force in 2006, have perhaps inhibited the development of policies and practices specifically to meet the needs of older workers, with organisations fearing accusations of age discrimination. By protecting this group on the one hand, it could be argued that we are putting them at risk on the other.

Age discrimination happens because assumptions are made about the older employees that are based on outdated, inaccurate and inappropriate stereotypes. Watson et al (2003) found some employers admitted they preferred to employ younger nurses; and policies including return to practice initiatives were not designed for those in their 50s. On the other hand, some employers voiced positive views of older nurses, valuing their courtesy, commitment and understanding of patients' needs. Furthermore, Duffin (2004) suggests that a lack of confidence in older nurses alters their perception of what they themselves can achieve.

Malone (2003) suggests that older nurses have gained valuable experience and expertise during their careers, and therefore suggests they should be valued whilst still on the payroll. Furthermore, research outside of the NHS indicates that there is a changing mindset of employers, who are beginning to recognise the value of older workers. Watson (2004) describes how current research suggests that the notion of the older work force being more unreliable should be dispelled, and goes on to suggest that the older work force are a valuable resource, demonstrating a successful working career and are in fact, good role models for their younger colleagues.

There would appear to be a more proactive approach to the ageing workforce and associated risk outside of the NHS. The Department for Work and Pensions (2001) found many areas of good practice in the recruitment and retention of older workers, highlighting that the effort and cost required in making adjustments to the ergonomics and nature of job roles, was far outweighed by the benefits of retaining the older workers. Benefits were highlighted as:

- High retention rates
- Lower absenteeism
- Reliability, commitment and dedication
- Developed people orientated skills and people development skills

- Flexibility and innovation
- Leadership and mentorship skills

Examples of good practice in employing older workers in private industry are:

- The DIY chain B&Q - normal retirement age is still 60, but employees can be given the opportunity to continue on fixed term contracts.
- Some engineering manufacturers have adapted equipment to meet the needs of older workers and are encouraging older workers to complete General National Vocational Qualifications (GNVQs).
- Hairnet is a company that provides computer training. Most of the trainers are over 50 and they specifically market themselves at potential learners who are over 50.

The economic driver of private sector business and the importance of maintaining their position in the market are the impetus for making the best use of their most expensive and valuable asset, the workforce. The NHS, however, appears to be slower at responding to this challenge and addressing this risk.

By the same token, it is crucial to establish what influences the nurse's decision to leave or stay. Watson et al (2003) found that positive reasons for staying in practice included:

- The opportunity to top up their pension
- Availability of flexible working options
- Continuing professional development

Reasons for taking early retirement included:

- Long term stress
- Difficulties in keeping up with technological changes

The government's proposed shake up of public service pensions, including the most controversial and wide ranging review of the NHS pension scheme since it began in 1948 (NHS Employers, 2005), may increase the risks older nurses pose to the NHS, both in terms of recruitment and retention and risks to the older nurses themselves. Harrison (2005) argues that the government are attempting to address the risk of not having enough nurses to meet the heath care demands of the population by increasing the retirement age to 65, whilst failing to recognise the emotional and physical demands of nursing which could prevent many nurses from being able to work up to that age. Moreover, Harrison (2005) points out that forcing nurses to work longer will lead to an increase in ill health retirement and end up costing the NHS more in the long term. Therefore, according to NHS Employers (2005), the compulsory rise in pension age will have unintended risks which will undermine the aim of retaining older staff, unless a range of proposed measures are implemented to support the retention of the older workforce such as:

- Job redesign
- Appropriate occupational health services
- Older care policies
- Provision of continuous professional development
- Tackling age discrimination practices
- Addressing the environmental pressures to ensure a healthy work-life balance

Watson et al (2003) found that the availability of flexible working practices is a key driver to retaining the older nurse, many of whom have caring responsibilities for elderly relatives. Part time working is the preferred option for this group, and this preference has a significant impact on the available workforce and increases the risk that supply will not meet demand. Many flexible working arrangements are geared towards younger staff, although the carer demands of older nurses are often more complex and demanding.

Altering shift systems to reduce the physical demands of the job may help accommodate and retain some older workers and reduce the risks to their own health, and their ability to do their jobs without putting patients at risk. White (1997) offers the view that older workers should work morning shifts as they are more likely to be productive and efficient, suggesting that people over 47 years are less able to adapt to the disruptive sleep patterns and disturbed circadian rhythms which are associated with shift work and could therefore increase the risk of making mistakes.

This view is supported by Reid and Dawson (2001) who found that when exposed to 12 hour shift rotations, older workers' (mean age 47) performance significantly decreased across the night shift compared with that of younger workers (mean age 21). There are, therefore, potentially significant consequences for older workers forced to undertake a 12 hour rotating shift pattern and night working, as it increases the risks of under performing and errors. Older nurses may also pose a risk to themselves and the patients if they are not physically able to perform the tasks required. Watson et al (2003) raise the question of older nurses being able to meet the physical demands, workload and stress associated with nursing, especially in areas with hi-tech requirements (e.g. intensive care or theatres).

Workability

The Royal College of Physicians, Faculty of Occupational Medicine (2004) describe the term 'workability' as *"the ability of workers to perform their jobs, taking into account the specific work demands, individual health conditions and mental health resources"* and go on to describe the ability to work as a function of:

- Health and functional capacities (physical, mental, social)
- Education and competencies
- Values and attitudes
- Motivation
- Work demands
- Work community and management
- Work environment

They conclude that at an individual level, these processes include promotion of good health and associated functions capacity, as these are pre-requisites for prolonging the ability to work effectively. Adequate training ensures competence to do the job and attention to work load, shift patterns and flexible working will avoid overwork and exhaustion. Therefore, pre-employment assessments and ongoing health monitoring are as important for the older workers as for any other group, especially as the older nurse heads towards retirement. A proactive approach to assessing nurses "workability" could prolong their ability to remain in post.

Lifting and handling (manual handling)

The Health and Safety Executive suggest that the lifting tasks of the nursing profession are comparable with those of the hardest labour. Therefore, the affects of such a physically demanding job could be identified as a risk to the older nurses in terms of manual handling injury. This would put patients at risk, especially if the older nurses were using outdated lifting techniques.

Manual handling accidents and injuries are not new. Guidance on how to make manual handling less hazardous has been available for years, but the problem has not gone away. Although training contributes to improving the situation, it is only one element of a successful approach which should also include risk assessment, ergonomic interventions and effective management policies.

Despite the introduction of the Manual Handling Operations Regulations (1992), back pain amongst nurses remains a problem. Gould (1998) suggests that nurses over 55 are 13% more likely to experience back injury. Furthermore, older nurses may have more of a stoical attitude to manual handling, with years of previous experience carrying out manoeuvres that have since been proven to be harmful to both the nurses and patients. Years of putting strain on their backs may have had a cumulative effect, making them more susceptible to injury. Conversely, Hollingdale (1997) found that a higher proportion of younger, less experienced nurses experienced back pain compared with the older more experienced nurses. Hollingdale (1997) suggests that the older population of nurses are the "survivor generation" with strong backs, supporting previous research suggesting that after age 50, the spine becomes stiffer; actually reducing the onset of back pain.

Current figures from the HSE indicate that 60% of all manual handling accidents reported (over 14,000 a year) involve patient handling and recent estimates suggest that as many as 3,600 nursing staff have to retire because of their injuries each year. This is despite high profile campaigns by the Royal College of Nursing and UNISON and the introduction of lifting equipment in most NHS hospitals. It would appear, therefore, that the risks associated with manual handling injuries are not attributable to the older nurses in isolation. Effective induction, training, ongoing support and the right equipment would seem to be the control measure required to protect the patients and the nursing population as a whole.

Stress

If the needs of the older workforce are not met, stress will likely be major risk factor for this group. Occupational stress amongst the nursing population in general is acknowledged as an area for concern. The Sainsbury Centre (2000) found that the percentage of the general working population with mental health problems is 17%, whilst for nurses it is 28.4%. Watson et al (2003) found that stress associated with burn-out were major influences on decision making over the age of 50. Ironically, staff shortages were seen as a major cause of stress that could potentially be alleviated by retaining these older nurses, or encouraging them to return. Watson et al (2003) also concluded that little had been done to combat older nurses' stress. Measures to reduce stress amongst older nurses, although effective when taken, had not been widely implemented.

The morale of older nurses is directly linked to their perceptions of being valued and levels of stress. Improving the morale of older nurses will encourage them to stay in the profession and encourage returnees. The climate of constant organisational change can make nurses who trained many years ago feel they have nothing in common with the nurses being trained today. Quant (2000) argues that older nurses may have been taught practices that are now out of date and proven through research to be no longer effective, and may therefore be putting patients and colleagues at risk. However, Quant acknowledges that the older style nurse training contained far more clinical experience and practical learning than today's training models. This view is supported by Benner (1984), who believed that previous learning and nursing experience provide an invaluable mix of theory and practice which is deeply embedded in the memory and rarely forgotten.

Clinical governance and training

Continuing to use outdated practice, whether intentional or unintentional, will put patients at risk. Older nurses should be actively encouraged to engage in clinical governance activities, such as reviews of clinical incidents, clinical audit activity and professional development reviews, to ensure their practice is current and any outstanding training issues are addressed. The implementation of Agenda for Change will force older nurses to engage with life-long learning if they are to progress through the key skills framework and realise their full earnings potential. Nurses will no longer be able to stagnate within their roles "doing what they have always done" However, this will come at a cost to the organisation both in terms of human resources investment required to manage the process and in meeting the training and development needs identified. Therefore, by controlling the risk of not training, organisations incur a financial risk of not meeting financial balance due to the burden of ever increasing training costs.

Increasing complexity

Taylor (2000) raises the issue of older nurses being able to cope with a complex and challenging work environment involving the need for increasing technological knowledge, highlighting that an incompetent workforce will increase the risks to patients. Quant (2001) supports this argument by suggesting that nurse returnees are often from a diverse age group and their needs will be equally diverse. Adults who are in the process of change either in their professional or personal lives might experience feelings of low confidence in their abilities, often viewing themselves as "out of date" and doubting their ability to learn new material. There is also evidence that older nurses find it difficult to cope with the pace of technological change and feel uncomfortable with the amount of high tech equipment found in clinical areas today (Duffin, 2004).

Malone (2003), however, challenges this view and illustrates society's growing acceptance of information technology both in the home and at work by the number of older people who surf the internet and exchange text messages with their grandchildren. Moreover, Taylor and Walker (1998) demonstrated no correlation with working practices of older workers and the ability to adapt to new technology and their interest in technological change; and argue that these findings need to challenge stereotypical attitudes towards older workers.

Therefore, if encouraging older nurses to return to practice is a major solution to the recruitment problem, it is important that "return to nursing" courses are planned effectively to meet the needs of the older learners and ensure they are competent within the clinical setting in order to minimise the risk to patients. However, it should be recognised that the pool of potential nurse returnees from which the NHS and other employers attempt to recruit is declining in numbers, as it also ages (Buchan, 1998).

Watson (2004) claims that older nurses returning following career breaks are more at risk, as their particular needs are not catered for on the courses. Much has changed in the NHS in recent years and nurses returning after significant career breaks may not be familiar with the changes. If return to practice courses do not address these particular needs, older nurses will be disadvantaged and more at risk in terms of patient safety and personal safety than a younger nurse returning after a shorter period. This approach could, however, be criticised for being ageist and generalising for all older nurses. Robust assessments of competence would need to be employed across the board for all returning nurses to ensure a non-discriminatory approach.

There is evidence to suggest that the physical and physiological affects of ageing can in turn affect learning. This will increase the risk of older nurses not being adequately trained to do the skills and tasks required and, therefore, putting patients at risk. Research suggests older workers learn at a slower pace and often their own perception of ageing can inhibit their ability to learn (Kiger, 2004). Furthermore, previous research into the affects of ageing on learning undertaken by Wellford (1962) as cited in Reid and Barrington (1999), suggests that ageing impairs the central decision making processes, this affects the time required to reorganise information, monitor movements and multitask. Wellford (1962) also found that age impairs short term memory, resulting in time increases and errors completing complex cognitive tasks.

Bromley (cited in Reid and Barrington, 1999) suggests that age is not the only factor to affect achievement and learning in later years. The individual's original intelligence can have an impact and the level of stimulation is important. Therefore, it can be argued that individuals who embarked on their

Registered Nurse training some years ago must have had a certain level of intelligence. Also, working within the health service is unquestionably challenging and stimulating, supporting the argument that this will increase mental ability. Moreover, the manipulative, occupational, mental and social skills acquired through experience help offset the decline in abilities as a result of the ageing process. Consequently, older nurses should be no less able to learn new skills and cope with the requirements of continuous professional development than any other age group. It is the techniques used to train older nurses which should be considered, to minimise the risks of an untrained or unskilled older work force. In support of this, Watson et al (2003) found that professional development and return to practice courses took little account of these issues.

The role and scope of nursing practice has changed dramatically over the past fifteen years, with a shift towards undertaking tasks previously carried out by doctors. This shift needs to be underpinned with support and training to ensure nurses have the right skills for the job. Without this, nurses run the risk of breaching their "Code of Professional Conduct" and being struck off the register.

Duffin (2004) states that nurses over the age of 40 are more likely to be struck off the professional register than younger nurses. Almost 80% of those removed from the register during 2004 were over 40. Duffin further states that the average age for nurses being struck off the register is 47. In contrast, only 2.45% of people in their 20s were struck off. Duffin suggests that the high numbers of older nurses finding themselves before misconduct hearings is a reflection of the changing nature of their work as they get older, and goes on to describe how some older nurses leave fast paced jobs and go into what they consider to be an 'easier' option, such as nursing home shifts or temporary work on agencies and nurse banks. This type of work can bring vulnerability and isolation, which in turn can increase the levels of stress which older nurses are seeking to relieve. Therefore, Duffin suggests that the transition to an alternative type of work at a later stage of a nurse's career is often more difficult than anticipated and brings about its own risks. Interestingly, although the figures suggest older nurses are more likely to be struck off the professional register, there is little empirical research to explain this phenomenon other than anecdotal evidence.

Concluding comments

The ageing nursing workforce is a global phenomenon which is growing, as an ageing population increasingly demands healthcare from a reducing and ageing nursing workforce. Steps to solve the problem by recruiting from overseas have redistributed the problem to countries less able to deal with the problem (Buchan, 2002).

The lack of empirical research into the effects and risks of the older workforce could be argued to be a risk in itself. Without this knowledge, we may never truly appreciate what the risks of an ageing nursing workforce are in order to address them, and ensure the continued safety of our patients, staff and the NHS as a whole.

The risk of a staffing crisis in the NHS is set to deepen as a predicted 15% of the workforce will be eligible for retirement within the next decade (White, 2002). The direct impact of these retirement figures is likely to be felt first by non-NHS sectors and by the NHS community nursing sector, but there will be a knock-on effect to NHS acute sectors. As a result, the sectors with relatively younger age profiles, for example teaching hospitals, might face increasing pressure to retain their staff. Employers will also have to note that the potential for skill substitution will be constrained by the high level of nursing auxiliary retirements. Furthermore, if the NHS is going to fulfil all of its plans for the modernisation agenda, it cannot afford to write off its older nurses (Malone, 2003).

Despite the government's efforts to stem the loss of nurses from the NHS, the literature suggests that older nurses frequently lack clear advice or guidance about their employment or retirement options. Little attention has been given to the scope for more flexible work and pension arrangements that would encourage nurses over 50 to stay in practice (Watson et al, 2003).

These findings show clearly that the NHS has not been devoting specific attention to nurturing older nurses as a valuable resource. The result is that nurses feel that they are not being helped by their employers to address difficulties associated with growing older in a rapidly changing service. This undoubtedly contributes to an earlier exit from nursing than might otherwise be the case. Yet there is much potential to reduce these difficulties and to make nursing more attractive, by focusing on the particular needs of older nurses, including those returning to the profession.

The published literature does not appear to support the assumption that work performance declines with age. Indeed, it is suggested that chronological age is a weak predictor of capacity for production or performance (Letvak 2002). Older workers are noted to perform generally more consistently and to deliver higher quality, matching the performance of younger workers. It can be argued that although there is an age related decline in physical strength, stamina, memory and information processing, this rarely impacts on work performance. Older workers often use knowledge, skills, experience, motivation and other strategies to maintain their performance. Older workers also bring the benefits of being more conscientious, loyal and reliable and hard working. And they tend to have well developed interpersonal skills. Research also suggests that older nurses recognise their need for continuing professional development to keep up to date with the rapidly changing field of health care (Malone, 2003). The government, however, appears to be sending mixed messages by identifying and acknowledging the risks of an older workforce, in terms of recruitment and retention, on the one hand, whilst on the other hand failing this group of nurses through encouraging a culture where flexible working and professional development are not seen as compatible (Duffin, 2004).

Although older workers may be perceived as less adaptable or able to accept change, training and support designed to meet the specific needs of the older workers can overcome this. Smarter working, using health promoting workplaces that provide flexible working, sensible work schedules, appropriately trained managers and workers, good practice in human resources management and proactive occupational health services, enable effective working for older workers. Older workers have the right to work safely; badly designed working conditions and a lack of training are the main factors that turn ageing into a risk rather than the ageing process itself. Watson et al (2003) found that there is a significant difference between the rhetoric of policy at government level and the actuality of policy at NHS level regarding the needs of older nurses.

To minimise any perceived or actual risks associated with an ageing workforce, the NHS needs to learn from the independent sector where a clear message has emerged that, where an industry does face particular recruitment and skill difficulties, there appears to be a greater willingness to recognise and address age diversity issues.

Finally, more detailed research into the needs of older nurses is urgently required to help the NHS to develop:

- Policies and practices that support older nurses in preparing for retirement.
- Family friendly working practices for those with caring responsibilities for older people.
- Continuous professional development tailored to the needs of the older learners to support training to update skills.
- Policies and practices that support older workers, including reviewing their job and work design to take account of their needs, particularly

physical working conditions, with regular occupational health reviews after 50.
- Return to practice programmes to address the unique needs of older nurse.

References

Benner P (1984). *From Novice to Expert: Excellence and Power in Clinical Nursing Practice.* California. Addison-Wesley.

Buchan J (1998). Nurse 'til you drop. *Nursing Standard* Vol 13 (15) p 34-35.

Buchan J (1999). The Greying of the United Kingdom Workforce: Implications for employment Policy and practice. *Journal of Advanced Nursing* 30 (4) p 818-866.

Buchan J (2002). Global nursing shortages. *BMJ* Vol 324 p 751-752.

Department of Health (1999). *Code of Practice for Age Diversity in Employment.* London. HMSO.

Department for Work and Pensions (2001). *Good Practice in the Recruitment and Retention of Older Workers: Summary.* London HMSO.

Duffin C (2004). Over 35's believe they are missing out on promotion. *Nursing Standard* Vol 18 (18) p 6.

Duffin C (2005). Over 40's struck off more often than younger staff. *Nursing Standard* Vol19 (23) p 5.

Grey J (2003). The needs of older nurses. *Nursing Standard* 29 Vol 18 (7) p 3.

Gould M (2000). New brooms and old hands. *Nursing Times* Vol 96 (35) p 6-8.

Harrison S (2005). Unions fight rise in pension age. *Nursing Standard* Vol 96 (35) p 6-8.

Health and Safety Executive (1992). *Manual Handling Operations Regulations.* London. HMSO.

Hollingdale R (1997). Back pain in nursing and associated factors: A study. *Nursing Standard* Vol 11(39) pg 35-38.

Ilmarinen J (2001). Ageing workers. *Occupational and Environmental Medicine* (58) 546-552.

Kiger A (2004). *Teaching for Health – Third Edition.* London: Churchill and Livingstone.

Letvak S (2002). Retaining the Older Nurse. *Journal of Nurse Administration* Vol 32 (7) pg 387-391.

Malone B (2003). Experience of a lifetime. *Nursing standard* No23 (18) pg 24.

National Audit Office (2004). *Welfare to Work: tackling the barriers to the employment of older people.* HC 1026 2003-2004. www.nao.gov.uk

NHS Employers (2005). *The NHS Pension Scheme Review: consultation.* London. HMSO.

Nursing and Midwifery Council (2004). *Statistical Analysis of the Register 1st April 2003 – 31st March 2004.* www.nmc-uk.org

Phillips P (2002). *Juggling work and care: The experiences of working carers of older adults.* Joseph Rowntree Foundation. Policy Press.

Quant T (2001). Education for nurses returning to practice. *Nursing Standard* Vol 15 (17) pg 39-41.

Reid M and Barrington H (1999). *Training Interventions: promoting learning opportunities.* Sixth Edition. London. Chartered Institute of Personnel and Development.

Reid K and Dawson D (2001). Comparing performance in a simulated 12hr shift rotation in young and old subjects. *BMJ* 58 (1) 58-62.

Roberts G (2002). *Risk Management in Healthcare.* 2nd Edition. London. Witherby and Co.

Royal College of Physicians (2004). *Faculty of Occupational Medicine: Position paper on age and employment.* London. HMSO.

Sainsbury Centre for Mental Health (2000). *Finding and keeping: review of recruitment and retention in the mental health workforce.* London. Sainsbury Centre for Mental health.

Taylor B (2000). *Reflective practice: A guide for nurses and midwives.* Buckingham. Open University Press.

Taylor P and Walker A (1998). Employers and older workers: Attitudes and working practices. *Ageing and Society* 18 pg 641-658.

Watson R et al (2003). *Nurses over 50: Options, decisions and outcomes.* Joseph Rowntree Foundation. Policy Press.

Watson R (2004). Older community nurses: Perspectives and prospects. *British Journal of Community Nursing* Vol 9 (7) pg 274-280.

White C (1997). Older shift workers should work morning shifts. *BMJ* 315: 1035-1038.

White C (2002). Ageing workforce will exacerbate NHS staffing crisis. *BMJ* Vol 325 page 1382.

This page is intentionally blank.

11

Violence and aggression towards health care staff - I

CAROLE MODERATE

Introduction

In today's NHS can be difficult for healthcare staff to develop meaningful relationships with their patients, and there is evidence that staff are being exposed to increasing numbers of angry patients, visitors and relatives (Winstanley and Whittington, 2004). The culture of the NHS is one of tolerance, with the emphasis on understanding and empathy and there is a degree of acceptance of behaviors which outside the healthcare environment would be seen as unacceptable. Stress and anxiety are common emotions in the healthcare setting, and these emotions can often progress to anger and manifest in aggressive and violent behavior (Hollinworth et al, 2005).

The Health and Safety Executive (HSE) define 'violence', including verbal abuse or threats as well as physical attacks, *'Any incident in which a person is abused, threatened or assaulted in circumstances relating to their work'* (Health and Safety Executive).

There are a range of data sources available on the extent of violence at work and its consequences in Britain, each defines violence in a different way, and no other source covers the entire range of violent incidents included within HSE's definition. The historical lack of a definition of violence has led to difficulties in comparing research data concerning frequency and incidence of violence and aggression in the NHS therefore raising the issues of data reliability and validity.

More recently, the NHS Counter Fraud and Security Management Service (CFSMS), launched as a special health authority in 2003, have produced their own definition of physical and non physical assault as:

Physical assault –
"The intentional application of force to the person of another, without lawful justification, resulting in physical injury or personal discomfort"

Non physical injury –
"The use of inappropriate words or behavior causing distress and constituting harassment"

Ferns and Chojnacka (2005) say that the word "intentional" in the above definition of physical assault causes a dilemma for healthcare staff who may have been injured by confused, demented or hypoxic patients. They go on to suggest that nurses are more likely to encounter unintentional rather than intentional violence. Lee (2001) supports this argument, claiming that the risk of violence amongst healthcare professionals is often underestimated because there is a failure to report cases where the assailant is not considered to be fully responsible for their actions.

Aggression and violence towards healthcare staff has become the focus of the UK government attention over recent years in an attempt to change attitudes and responses to such behaviors. The two main sources of data on the British working population are the British Crime Survey (2002/2003) and reports made to the HSE or local authorities under the Reporting of Injuries, Disease and Dangerous Occurrences Regulations (RIDDOR, 1995). The CFSMS have also developed a mandatory national reporting system which requires an investigation of all physical assaults on NHS staff.

Comparing the data from all three sources will identify trends in reported violence. However, these statistics alone may not be the full picture as it is important to consider the human context in which violence occurs to understand why healthcare staff are apparently experiencing an increase in workplace violence. The culture in which healthcare staff report incidents should be explored along with the culture of management in response to incident reporting, investigation and their willingness to take actions.

Violence in the NHS

The problem of violence in the NHS is not new, what has changed is the degree of interest in the phenomenon, which has in part been increased by a number of official reports by national bodies regarding workplace violence from the late 1980's through the 1990's, the consensus of opinion from the early studies was that violence in healthcare is a very real and growing problem, although previous evidence suggests a long history of under reporting (Rew and Ferns 2005).

The Health Service Advisory Committee (HSAC, 1987) found that 34% of nurses were attacked on duty, and half of these incidents occurred in A&E departments. Further studies by the HSAC (1997) suggested that 7% of all workers had been subjected to physical assault in the workplace, whilst the figures for nursing was 34%, this being the largest percentage for any occupational group. However, hospitals as places of work may be over represented in terms of levels of notifiable injury resulting from assault.

From a review of the historical studies on healthcare violence, it would seem that nurses have always been at higher risk of assault than workers in general. However, the available literature is limited and the comparisons between data studies should be viewed with caution as there are limitations to the data, explained by Cole (2003) as:

- The interchangeable use of the terms "assault" with "abuse"
- The vast differences in the definition of "violence" some studies including assaults with injury only ad excluding assaults where no injury has been sustained.
- The time periods for data collection varied as did the way in which data was collected, i.e. targeting staff following repeated assaults or those with a known injury, may skew the results and lead to reporter bias.
- Lack of formal incident reporting systems and underreporting also make the data unreliable.

Cole points out that the lack of valid comparative data means the questions the government's ability to claim a success of the Zero tolerance Campaign (Department of Health, 1999), which required all Trust to cut levels of violence and aggression against staff by 30% by 2003.

Most of the early studies of violence towards NHS staff were undertaken within the psychiatric and A&E settings (Fottrell, 1980; Cardwell, 1984). It has long been acknowledged that the accuracy of such studies relies heavily on the reporting behaviours of the staff therefore caution should be given to long term historical comparisons of data.

Wells and Bowers (2002) agree that much of the early research into NHS violence originates from the psychiatric settings and because of the vastly differing nature of the patients, care settings and approaches to care, generalisation with acute settings should be treated with caution as they may give an inaccurate picture. In addition, studies which look at violence towards nurses alone fail to consider other NHS professional and therefore do not offer a true picture of the incidence of NHS violence as a whole.

Winstanley and Whittington (2004) agree that the main body of research around violence and aggression has historically concentrated on psychiatric settings, they go on to comment that very few have considered general hospitals or looked at the variations between professions and locations, and argue that institutional averages actually obscure actual levels of violence experienced by particular professions or departments. To illustrate this point Whittington et al (1996) found that 22% of radiographers, 19% of doctors and 17% of physiotherapists reported being assaulted in that preceding year.

Government implemented measures to address violence in the NHS are also not new, Cardwel (1984) noted that the Department of Health and Social Security published guidelines for action following HSE studies in 1976, but that few Health Authorities had implemented them by 1984. Cardwell (1984) suggests this indicates that unless guidelines are mandatory and monitored they have very little effect on the working environment.

Violence - prevalence and location

Violence is inherent within our society and it is inevitable that violence will be part of NHS life, and the level of violence will reflect the community in which that service is provided (Scott, 2003 and Atawneh et al, 2003). In 2003 the HSE reported that assault had become the biggest cause for the over 3 day injuries to healthcare workers. However, the British Crime Survey found that, overall, the number of incidents involving violence at work across all occupations had actually fallen by 35% from a peak of 1,310,000 in 1995 to 849,000 during the year reported.

According to the HSE, trends in violence at work are difficult to interpret, although there is some evidence to suggest numbers of cases have remained relatively stable in the last few years, with some occupations being at greater risk than others; the major occupational groupings with the highest risk of assault were protective service occupations at 12,600 per 100,000 workers, followed by health and social welfare associate professionals at 3,300 per 100,000 workers (British Crime Survey, 2002/03).

The Healthcare Commission (2004) found from the 2004 NHS staff survey that there was little change since 2003 in the levels of violence and harassment reported by staff. Some 27% of staff had been harassed, bullied or abused at work in the past twelve months by patients or their relatives; this rises to 37% if bullying and harassment from colleagues is included. 14% of respondents had been physically attacked by patients or their relatives in the past year, and a further 1% of staff reported experiencing violence from colleagues. Data from the author's employing Trust (Table 11.1) demonstrates a significant increase in reporting of violence and harassment across the Trust over the two years to March 2005. This local phenomenon is thought to be a direct consequence of receiving an improvement notice from the HSE for the management of violence and aggression on the 14th October 2003. This followed investigation into a violent incident where a member of staff was assaulted by a patient in April of the same year.

Cause Group: violence and harassment- (Total reported incidents)	
April 2003-March 2004	210
April 2004-March 2005	311

Table 11.1 - Total reported incidents involving violence and harassment to staff

It should be noted that during 2004/2005, personal safety training was introduced within the Trust as part of the action plan towards meeting the requirements of the HSE Improvement Notice and to fulfill the requirements for training laid down by the CSFMS. This led to an increase in awareness by staff of the need to report all violent incidents. The cause codes within the cause group of violence and harassment were extended during 2004-2005 to give more explicit information on the type of violent incidents being reported. This also increased staff awareness regarding the nature of incidents to report, possibly leading to an increase in reporting. Winstanley and Whittington (2004) warn that expanding the definitions will increase figures of occurrence and this should be considered when comparing data.

There is an assumption from literature reviewed that most violence and aggression occurs in the A&E department. However, Whittington et al (1996) found that 90% of reported incidents for violence and aggression took place outside of A&E. Moreover, Winstanley and Whittington (2004) also suggest that this assumption is misplaced, quoting studies from as early as 1987 that found that staff in medical wards were more at risk than A&E staff.

Data from the author's employing Trust (Figure 11.1) supports the previous findings, with general wards being higher reporters of violent incidents than A&E, this trend may be attributable to the shorter waiting times, (Hinchingbrooke is one of the best performing Trusts in the country for the 4hr target in A&E according to 2005 Healthcare Commission data). It appears that high levels of aggressive behavior are no longer seen in the traditional A&E environment, but these behaviors continue to occur in the onward receiving wards.

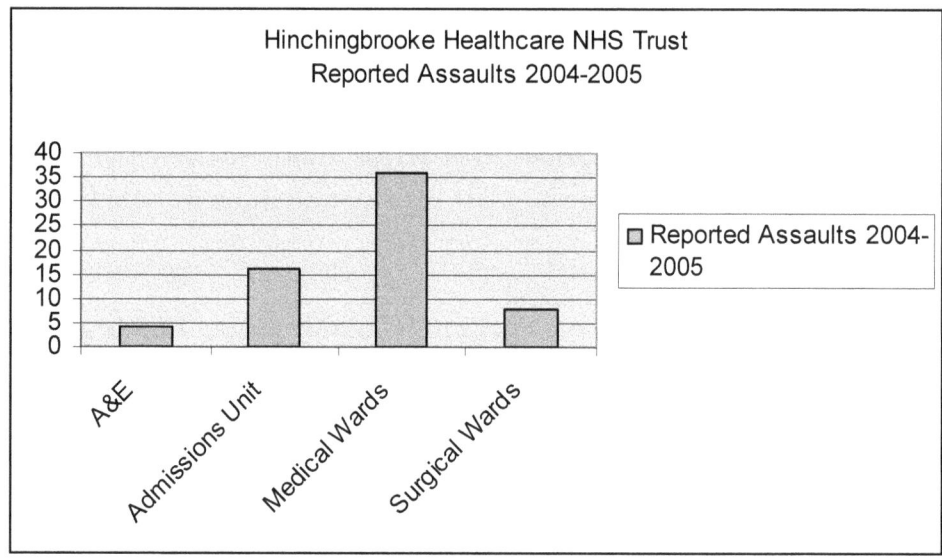

Figure 11.1 - Hinchingbrooke Healthcare NHS Trust: Reported Assaults (2004-2005)

Although the literature reviewed suggests that A&E may not be a particular focus for violent assaults, evidence suggests that A&E staff do suffer from high levels of threatening behaviours, which may often go unreported as staff become desensitised to the effects of such behaviour. In comparison, whilst women's services and children's wards often experience the lowest levels of aggression from patients, they have a high incidence of aggression from relatives and visitors, possibly because patients in these areas could be perceived to be more vulnerable and in need of protection (Winstanley and Whittington, 2004). Again, this trend is reflected in the recorded incidents of verbal abuse at the author's employing Trust (Figure 11.2).

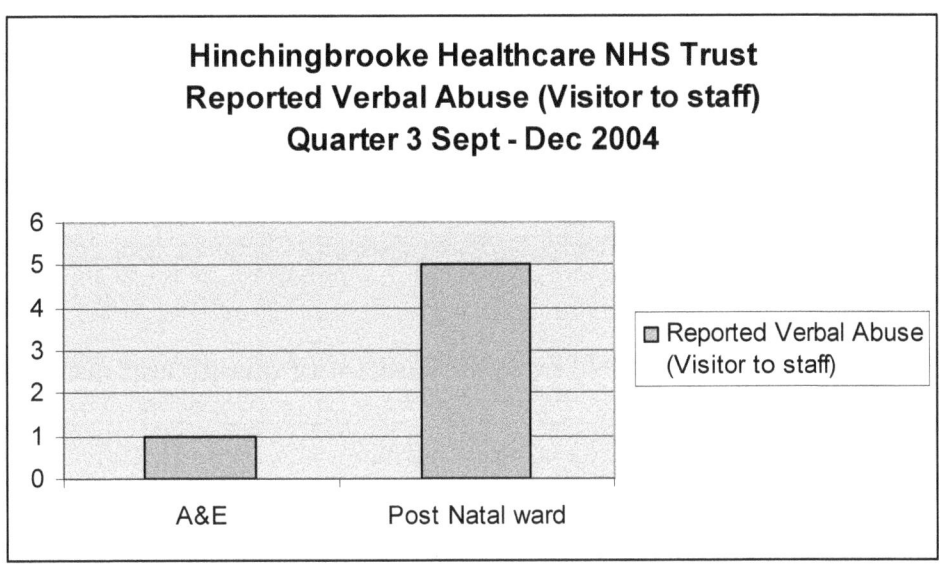

Figure 11.2 - Hinchingbrooke Healthcare NHS Trust : Reported verbal abuse (visitor to staff)

Cole (2003) comments on the fact that most literature on healthcare violence and aggression has focused on acute care settings, and goes on to argue that the problem is arguably more worrying in primary care, where incidents are frequent and backup is often less predictable with a higher number of vulnerable lone workers. Hollinworth et al (2005) agree, and suggest that often feelings of anger initially expressed in secondary care can stay with the patient and continue to be expressed to primary care practitioners. Cole (2003) goes on to point out that community workers are more vulnerable because they are on someone else's territory, therefore promoting reporting, risk assessments and control measures to protect community staff are very important.

Media influence

It could be argued that the media reporting of NHS violence exacerbates the problem. A sensational headline in a tabloid news paper may sell copy, but could present an unbalanced approach to the problem, raising the perception of fear beyond the reality. Wells and Bowers (2002) observes that the media usually claim large numbers and use descriptive language which intimates violence is a huge problem for the NHS, which could adversely affect recruitment and retention.

For example, on 14 June 2005 the Norwich Evening News claimed that Norwich has been named as one of the worst hit cities for the number of violent attacks on healthcare staff, carrying the headline "Attacks on Nurses Soaring " The article stated the following:

"The last official figures released by Norwich Primary Care Trust revealed there were 738 violent incidents against nurses, GPs and healthcare staff in 2003"

"Last year 139 members of staff including nurses, doctors, paramedics and clerical workers were attacked at the Norfolk and Norwich University Hospital by patients or visitors. The previous year there were 74 physical assaults between July 2002 and June 2003."

Comment included in the article from the director of quality and nursing at the Norfolk and Norwich Primary Care Trust indicated that staff training in the management of violence and aggression had been a priority in 2003, raising the awareness of staff and encouraging reporting. She went on to claim that all staff involved in the same incident are asked to report it, therefore, the numbers of reports are higher than numbers of actual incidents. Also, the numbers reported reflect verbal abuse incidents as well as physical assaults which is not made clear in the report. This highlights the importance of fully understanding what the numbers actually represent in terms of the real picture.

A further quote from the same article stated:

"And countrywide attacks on healthcare staff are soaring; with more than 300 assaults taking place each day amounting to 116,000 a year"

There is no reference to the source of these figures or the opportunity to validate them, and the public is led to believe that working for the NHS is very dangerous.

In the same context, television dramas such as "Casualty" (BBC 1) often portray healthcare workers facing violence and aggression, although these scenarios are often played out sympathetically towards the NHS staff, very rarely do you see the incident reporting process in action. Hollinworth et al (2005) supports this view and comments that the media and TV dramas can influence people's perceptions and behaviours around violence and aggression.

Causes and effects

The literature reviewed discusses different causes for violence towards healthcare staff. Hollinworth et al (2005) describes how government and management influences on the way care is delivered has lead to a more task orientated delivery of care (waiting list initiatives, 4 hour A&E targets, etc.), which gets in the way of a nurses attempt to provide a patient-centred approach to care. In turn, this makes it more difficult for healthcare staff to develop therapeutic relationships with patients which, when added to a patient's high expectations of the NHS, can result in an aggressive response to a problem becoming more likely. Hollinworth et al (2005) are also critical of the care pathway model of providing care, arguing that this approach means there are often many different healthcare professionals inputting into the care of one individual, thus giving the impression of fragmented and insensitive care leading to anger and frustration on the part of the patient or relative.

Atawneh et al (2003) describe how the effects of violence and aggression can differ depending on the severity and frequency of episodes and the perceived vulnerability to further episodes, pointing out that the knock-on effect to the other staff may be defensive patterns of behaviour that actually promote more violence and aggression. This view supports previous findings by Nabb (2000) who noted a direct relationship between violence and aggression and sick leave, increased alcohol intake and drug usage, and staff turnover. Nabb further implies that these behaviours fostered a negative attitude toward work by the whole team which results in higher stress levels and behaviours that may promote a violent response from patients. This view mirrors that of Cutcliffe (2001), who found evidence that, on occasions, some nurses provoke violence from patients.

Winstanley and Whittington (2004) considered contributing factors between those who experience no aggression and those who suffered repeated victimisation. Their study found a commonality between increase in violence and the extent to which staff are trying to influence patients' behaviours. Hollinworth et al (2005) suggests that it is possible that practitioners unintentionally compound these stressful situations, and observes that nurses often bear the brunt of patients' frustrations, anger and perceived loss of control. Nurses need to be able to recognise circumstances or changes in the emotional status of an individual that may trigger an angry response, and develop strategies to manage the angry individual in a professional and efficient way. Hollinworth et al (2005) goes on to comment that taking an authoritarian approach may increase the anger in someone who already feels they have lost control.

The cost of violence and aggression

The cost of violence and aggression in the workplace to an organisation can be considerable. The NAO (National Audit Office, 2003) estimates the annual cost to the NHS is at least £173 million. According to Rew and Ferns (2005), costs that can be incurred include:

- Sick pay if the staff member is injured
- Additional costs of temporary staff
- Counselling if needed
- Legal fees – staff injury claim
- Loss of experience and recruitment and training costs should the member of leave
- Cost of implementing controls Vs the cost of investing in staff
- Cost of potential HSE prosecution
- Loss of reputation

In human terms, the impact of a violent assault may be considerable. A direct physical assault can cause staff to suffer physical effects, although psychological effects can be equally disabling. The outpouring of emotion following violence is to be expected in the initial hours and days following an episode and can include:

- sleep disturbance
- tearfulness
- irritability
- loss of concentration
- episodes of hyperventilation

Some effects can be expected as a normal adjustment response. However, continuing symptoms affecting lifestyle require intervention and professional support (Winstanley and Whittington 2004). Serious and persistent verbal abuse may damage worker's health through anxiety and distress. In addition, worry about violence at work, even in workers who do not directly experience it, or the perception of the risk of violence can be a source of stress.

Estimates from the British Crime Survey (2002/03) indicated that 3000 per 100,000 workers (3%) were "very worried" about being assaulted at work and a further 9,000 per 100,000 (9%) were "fairly worried". In addition, 500 per 100,000 workers (0.5%) reported that worrying about violence at work had affected their health "a great deal," and a further 2000 per 100,000 workers (2%) said that worrying about violence at work had affected their health "quite a bit."

The effects of verbal abuse can also be damaging. Crouch (2004a) notes that the impact of verbal abuse depends on the nurse patient relationship, arguing that if the nurse has built a therapeutic relationship with the patient prior to the verbal abuse, then the effects are more damaging. However, it is precisely this therapeutic relationship which is thought to prevent abuse in the first place. Crouch (2004a) goes on to comment that there is evidence that specialities who are quick to report verbal abuse and take action to stop it can boost the nurses confidence and morale, which protects them from the psychological effects of abuse.

Reporting and Culture

Both employers and employees have obligations in respect of violence at work under the requirements of the *Health and Safety at Work etc. Act 1974*. Under the *Management of Health and safety at Work regulations 1999*, employers are required to consider the risks to employees, which include the risks of reasonably foreseeable violence, and decide how significant these risks are. Controls measures must be put in place to mitigate these risks and a clear management plan should be communicated to demonstrate how these risk management strategies will be achieved. This is only achievable if employers are made aware of the risk through staff reporting any and all untoward incidents. Travis (2005) suggests that NHS employers are failing in their duty of care to nurses and the NHS workforce because they do not tackle violence at source. A competent reporting process will enable the situation in general to be monitored and allow for investigation, root cause analysis and evaluation of initiatives in place to reduce violence and aggression.

Patterson et al (1999) comment that if violence is a foreseeable risk it should be seen as a key health and safety issue. Organisations that fail to comply with the legal requirements may receive improvement or, less likely, prohibition notices from the HSE. They may even face prosecution for failing to carry out proper risk assessments. The importance of reducing workplace violence and aggression is reflected in the Health and Safety Commission's Revitalising Health and Safety Strategy Statement (2000) which contains a target that specifically focuses on reducing workplace violence and aggression in the NHS. The emphasis is on proactive risk assessments and effectively dealing with violent incidents once they have happened. Munro (2002) points out that under reporting makes it difficult for employers to meet their legal obligations to protect staff from danger at work, leading to a lack of incident investigation and no follow up risk assessments.

However, the NAO (National Audit Office, 2003) found that approximately two in five incidents go unreported, with doctors particularly reluctant to report. They also found under reporting to be a significant issue in other public service industries and suggest that the main reasons for not reporting violent incidents are:

- A poor reflection of an individual ability to cope manage the incident
- Not wanting the attention reporting will bring
- Forms not user friendly
- Culture of acceptance
- Lack of follow up from previous reported incidents
- Lack of support from colleagues

Wells and Bowers (2002) suggest that in terms of official records, substantial under reporting has been confirmed, which may relate to a belief amongst nurses that violence is just part of the job. Lehane and Carver (2003) agree, putting forward the concept that as all health care workers are potential victims of aggression, they themselves have a responsibility for ensuring violence is never taken lightly. However, Lehane and Carver (2003) also point out that it is often the nurses coping mechanisms and a desire to 'get on with the job' that are inhibitors for reporting violent episodes.

A reporting culture could be influenced by individual definitions. Also, according to Johnson (2004), perceptions of violence and aggression, cultural and past experiences can influence reporting behaviours. In contrast, Munro (2002) found that it was the actual severity of the incident that influenced reporting and that the nurse's gender and patient characteristics had no relevant effect on reporting.

Earlier studies undertaken by Beech (2001) found that staff groups who frequently work in speciality areas of health and social care such as mental health and older people, which encounter higher levels of violence and aggression, have an accepting approach to violence in the work place. The implication is that this culture of acceptance inhibits reporting as incidents of violence are mismanaged in a variety of ineffective ways:

- Denial
- Inappropriate humour
- Excessively withdraw
- Excuse violent behaviours towards new and inexperienced staff as a mythical "rites of passage"

Crouch (2004b) also talks about a culture of acceptance in mental health specialities, pointing out that twice as many mental health nurses as A&E nurses said a degree of abuse is "acceptable" (36% and 17%, respectively).

In the same way, Rew and Ferns (2005) refer to the effects of the "team culture", stating that a supportive, cohesive team approach encourages reporting. Equally, Munro (2002) argues that a poor team culture can adversely affect reporting as a "blame culture" emerges. A nurse's competencies may be called into question if an incident of violence occurs on their shift and reporting the incident draws attention to the nurse's perceived failure. Munro (2002) goes on to emphasise that this poor team culture actually exacerbates violence as unreported events lead to a lack of communication within the team. Munro further suggests that patients often remain volatile in the immediate aftermath of a violent incident and, consequently, a proportion of nurses may walk unprepared into a dangerous situation and inadvertently reignite the situation.

Munro (2002) also looked at the reasons nurses don't report violent incidents and what influences their reporting behaviours. Under reporting is often a symptom of a poor response from the organisation once incidents are reported, leading to little or no action being taken. This point is emphasised by Travis (2005) who reported that following a physical assault, no follow up actions were taken in nearly 80% of cases. Only 5% of cases were referred to the police and in only 2% of cases was the offender prosecuted.

Lehane and Carver (2003) found that even in organisations with a culture of high reporting, there was a general lack of response from management, with those reporting receiving little or no feedback. Eventually, this lack of feedback can discourage further reporting of subsequent incidents. It is all too easy for management to misinterpret this apathy to report, and be under the false belief that the problem of violence and aggression is under control and being managed effectively.

Wells and Bowers (2002) observe that once admitted to hospital, the removal of a violent patient becomes more difficult. Some Trusts have devised policies and procedures to "Red card" individuals, making it possible to refuse treatment and admission of identified violent individuals. This process is only workable if fully supported by commissioners and primary care stakeholders.

Kerr (2004) suggests that a collaborative approach between managers, occupational health, colleagues and organisational policy and strategy is the key to supporting and reassuring staff that it's ok to report incidents. Kerr (2004) goes on to emphasise that actions following reporting must be consistent to reinforce the message of a positive reporting culture.

Training

According to Scott (2003), despite the growing evidence from official reports on the increasing incidence of workplace violence, there is still reluctance by Trusts to engage with robust and effective risk assessments to protect staff, limited and variable quality of staff training and a lack of support or counselling for staff following reporting an incident. Scott believes that if violence is to be reduced within healthcare settings, a much greater emphasis and commitment to its reduction is required.

To reduce the organisation's risk of liability, Bleetman and Fayeye (2003) suggest that effective training for staff in the management and prevention of violence and aggression is the key. They believe that training needs to be focused on communication skills and customer care and should be a requirement for all NHS staff at all levels. Scott (2003) suggests that targeting senior staff, who are often seen as role models, is a cost effective way to influence a culture change. On the other hand, Taylor (2000) suggests that junior staff and students are at high risk from violence and argues the case for pre-registration training being mandatory in all schools of nursing.

More recently, the NHS SMS published a national syllabus for conflict resolution training for all NHS frontline staff in April 2004. The training gives staff the skills to recognise and defuse potentially violent situations. Separate syllabuses for those working in mental health and learning disabilities are being developed by a CFSMS-led group. They will also consider the issue of physical intervention techniques. At the time of writing, more than 85,000 frontline staff had been trained (Rew and Ferns 2005).

Bleetman and Fayeye (2003) argue that there is often a discrepancy between what is taught and what is needed, recognising that the danger of teaching physical breakaway techniques or even control and restraint may increase violent attacks on healthcare professionals as more individuals may be inclined to 'have a go'. Bleetman and Fayeye also point out that, in law, individuals have the right to self defence, but it must be reasonable and proportionate, and any Trust Policy which forbids staff to intervene physically in any conflict is unlawful as it prevents staff from exercising their right to self defence. On the other hand, section 44 of the Employment Rights Act (1996) states that;

"Staff has the right to withdraw from his or her care environment if he or she feels seriously under threat"

Therefore, each course of action is covered by law and staff need to be aware of their options through training and risk assessment to ensure the most appropriate action is taken in the event of a violent incident.

Rew and Ferns (2005) support the need for staff training at all levels but suggest that communication and de-escalation skills are the priority. They go on to imply that the need for breakaway techniques suggest that violence has already taken place, and that it focuses on a reactive management of the problem. The investment required in terms of time and funding to ensure breakaway techniques are leant correctly and embedded with regular practice to be safe for both the staff and the aggressors may be too high for organisations to ensure breakaway techniques are effective.

Lee (2001) acknowledges the importance of teaching nurses to recognise the cycle of violence, but suggests a more therapeutic approach to interrupting this cycle before the need for breakaway occurs. Lee talks about the theory of self efficiency, which is part of the general theory of behaviour change. Bandura (1977, cited in Lee 2001) talks about behaviour changes being directly linked with the modification of beliefs and expectancies. Self efficiency, therefore, is the belief that an individual can succeed at a given task or behaviour. It is not concerned with skills, but with judgement of what one can achieve. The theory argues that staff with high self efficacy regarding managing potentially violent situations will envisage a favourable outcome and can be more confident in dealing with the conflict, believing that they can undertake breakaway if required. Those with low self efficacy in their ability to manage a violent situation will envisage a less favourable outcome of possible harm and be less likely to succeed in the de-escalation. Lee suggests that the type of training given to nurses to manage violence and aggression should aim to increase the feelings of self efficacy if they are to be effective.

Management by law

The NAO (National Audit Office, 2003) states that there were only 50 identified prosecutions for assaults on NHS staff during the reporting year they considered in their report. They were critical of NHS organisations for the lack of action taken against offenders. It is anticipated that this situation will change with the introduction of the NHS Security Management Service (SMS) and the requirement that, by June 2006, a Local Security Management Specialist (LSMS) be based within every NHS organisation. In partnership with the police, the LSMS will investigate assaults on staff and ensure appropriate action is taken. Also, an NHS Legal Protection Unit was set up in December 2003 by the NHS SMS to ensure legal action is taken against anyone who assaults NHS staff.

Evidence of the change in the management of violent and aggressive patients can be seen in the case of Norman Hutchins who, in June 2004, became the first person ever to receive an NHS-wide Anti-Social Behaviour Order and was imprisoned for three years for harassing NHS staff (BBC News, 2004).

The police have no automatic powers to arrest someone who assaults a nurse at work. Common assault which results in "minor" injuries is not an arrestable offence, although individuals may be arrested for breach of the peace.

The NHS CFSMS is currently funding and pursuing the first private prosecution of someone who has assaulted a nurse and there are calls to make assaulting a nurse or any other public or emergency sector worker whilst engaged in providing a service to the public an aggravated assault, which already carries a harsher sentence (Parish and Waters, 2005). This is similar to a scheme already in existence for Scotland under *The Emergency Workers Act 2005 - Scotland*. But the NHS SMS Chief Executive argues that adding a specific offence of violence against healthcare workers will not reduce violent attacks (Choudhari, 2005). Trade Unions and professional bodies (RCN) are disappointed by the response of the NHS SMS and claim that more prosecutions will send the right message to staff and public alike, ultimately reducing the incidence of violence and aggression in the NHS. Nonetheless, Parish and Waters (2005) argue that despite any new proposed changes to the law, the real power to change the way assaults are dealt with rests with the Crown Prosecution Service (CPS) and the courts.

Conclusions

The NHS is Europe's largest employer, employing over 1 million staff (NAO 2003) and all NHS staff has a right to expect a safe and secure working environment. NHS organisations have a legal and ethical duty to prevent staff becoming harmed by violence and aggression by assessing and mitigating, where possible, the foreseeable risks.

Healthcare environments are alien to all but those who work there. Most people are fortunate enough never to consider the healthcare system unless or until they, or someone they know needs to access it. This is often during periods of anxiety and stress related to ill health, this heightened state of anxiety means that it does not take much to evoke an angry response.

Wells and Bowers (2002) question whether the growth of interest in NHS violence represents a response to an increasing problem, or simply the over due recognition of a long standing one. On the other hand, although there is a perception that violent incidents towards NHS staff is increasing, there is little evidence to support this, and that as the means of monitoring and assessing any changes are consistently changing, it is difficult to make accurate comparisons (Lehane and Carver, 2003).

The new mandatory national reporting system introduced by the CFSMS may help to address this issue, by providing a consistent way of gathering and reporting data on violent incidents. However, it may be some time before any useful comparative statistics are available. The success of the SMS mandatory approach to training will also require more evaluation at a later date, interestingly, the government has not thought it necessary to invest financially in Trusts to enable them to achieve this mandate; therefore, this activity will be added as another cost pressure to already financially constrained Trusts.

Reporting of violent incidents is crucial in monitoring their occurrence and assessing the success of measures and controls implemented to combat violence and aggression in healthcare. However, some evidence suggests that an increase in reporting of violence towards NHS staff merely reflects changing attitudes to violence in the wider society (Wells and Bowers 2002).

According to Kerr (2004), it is important to encourage reporting of violence and aggression through effective training, as failure to provide training and support for staff not only makes it difficult to estimate the true picture of actual violence, but it also reinforces the belief that workplace violence is somehow acceptable. With an effective and robust reporting system, the true picture of the scale of the problem can be identified. Effective strategies for dealing with the violence and aggression can be implemented and effectively evaluated. Individuals who pose a significant risk could be identified before a serious incident occurs.

It is clear that all healthcare staff need to have the training and self awareness to recognise the situations which have the potential to result in violence and aggression. Staff need to be able to react appropriately to angry outbursts without fuelling the situation using the skill of de-escalation and if necessary, breakaway. Healthcare staff should reflect critically on the types of relationships they develop with patients, using well developed communication skills to prevent an angry outburst escalating into violence. Although it should be acknowledged this is no easy task when the member of staff may be feeling frightened, vulnerable or unjustly criticised (Hollinworth, 2005).

Protecting NHS staff from violence and aggression has to be priority for all NHS organisations. However, staff can help themselves and their colleagues by always reporting, and documenting in the patients notes any violent incidents. It is important to communicate this information across the organisation and across organisational boundaries where the violent patient may move from one care setting to another, putting the receiving staff at risk if they are unaware of the patient's violent history.

Removing the blame culture will help healthcare staff realise that they do not have to accept violence as part of the job. Encouraging open and honest communication within an organisation regarding violence and aggression should be supported by management to include debriefing following incidents. This may be a costly investment in terms of both finance and time required, but ultimately, this investment in staff and their safety will have a positive effect on the culture of the organisation, encourage good reporting and effective management of violence and aggression (Munro 2002).

References

Atawneh F et al. (2003). Violence against nurses in hospitals: prevalence and effects. *British Journal of Nursing* Vol 12 (23) pg 102-107.

BBC News (2004). Man banned from every hospital. http://news.bbc.co.uk/2/hi/uk_news/england/north_yorkshire/3770081.stm

Beech B (2001). Zero tolerance of violence against healthcare staff. *Nursing Standard* Vol15 (16) pg 39-41.

British Crime Survey (2002/2003) www.homeoffice.gov.uk/rds/bcs1.html

Bleetman A and Fayeye O (2003). Preventing and Managing aggression and violence in the NHS. *Hospital Medicine* Vol 64 (12) pg 728-731.

Cardwell S (1984). Violence in Accident and Emergency Departments. *Nursing Times* 80 (14) pg 30-34.

Choudhari H (2005). Security advisors reject calls for new law protecting health workers from assault. *Nursing Standard* Vol 19 (40) pg 8-9.

Cole A (2003). Government health warning. *Health Service Journal* Sept 4 2003 p. 24-25.

Crouch D (2004a). Are employers tackling verbal abuse. *Nursing Times* Vol 100 (41) pg 24-25.

Crouch D (2004b). Abused specialities. *Nursing Times* Vol 100 (34) pg 24-25.

Cutcliffe J (2001). The answer to reducing patient violence towards NHS staff. *British Journal of Nursing* Vol 10922) pg 1442.

Department of Health (1999). *Zero Tolerance Zone Campaign*. Department of Health. London.

Employment Rights Act (1996). HMSO, London

Ferns and Chojnacka (2005). Reporting incidence of violence and aggression towards NHS Staff. *Nursing Standard* Vol 19 (938) pg 51-56.

Fottrell E (1980). A study of violent behaviour among patients in psychiatric hospitals. *British Journal of Psychiatry* 136 pg 216-221.

Healthcare Commission (2004). NHS staff survey.
www.healthcarecommission.org.uk

Health and Safety Executive. Definition of 'Violence'.
www.hse.gov.uk/violence/index.htm

Health Service Advisory Committee (1987). *Violence to staff in Health Services.* Health and Safety Commission, London. HMSO.

Health Service Advisory Committee (1997). *Violence to staff in Health Services.* Health Services Commission, London. HMSO.

Travis M (2005). Safer working. *Nursing Standard* Vol19 (28) pg21.

Health and Safety Commission (2000). Revitalising Health and Safety – Strategy Statement. www.hse.gov.uk/revitalising/strategy.pdf

Health and Safety at Work etc. Act (1974). HMSO, London.

Hollinworth H et al (2005). Understanding the arousal of anger: a patient-centered approach. *Nursing Standard* Vol 19 (37) pg 41-47.

Johnson N (2004). Strategies for dealing with aggression and violence. *Nursing and Residential Care.* Vol 6 (5) pg 237-239.

Kerr A (2004). "gonnae no dae that". *Nursing Standard* Vol 19 (3) pg 13.

Lee F (2001). Violence in A&E: The role of training and self efficacy. *Nursing Standard* Vol 15 (46) pg 33-38.

Lehane M and Carver L (2003). Hurt feelings. *Nursing Standard* Vol18 (10) pg 19.

Management of Health and Safety at Work Regulations (1999). HMSO, London.

Munro V (2002). Why do nurses neglect to report incidents? *Nursing Times* Vol 98 (17) pg 38-39.

Nabb D (2000). Visitor's violence: The serious effects of aggression on nurses and others. *Nursing Standard* 14 pg 36-38.

National Audit Office (2003). *A Safer Place to Work: protecting NHS Hospitals and Ambulance Staff from Violence and Aggression.* NAO, London. www.nao.gov.uk

Norwich Evening News (2005). Norwich Evening News. 14th June 2005

Parish C and Waters A (2005). Tougher sentencing is the key to a safer NHS. *Nursing Standard* Vol 19 (35) pg 14-16.

Patterson et al (1999). Violence at work. *Nursing Standard* Vol13 (12) pg 43-46.

Rew M and Ferns T (2005). A balanced approach to dealing with violence and aggression at work. *British Journal of Nursing* Vol 14(4) pg 227-232.

Scott H (2003). Violence against nurses and NHS staff is on the increase. *British Journal of Nursing* Vol12 (7) pg 396.

Taylor D (2000). Student preparation in managing violence and aggression. *Nursing Standard* Vol 14 pg 39-41.

Whittington et al (1996). Violence to staff in a general hospital setting. *Journal of Advanced Nursing* 24 pg 326-333.

Winstanley S and Whittington R (2004). Aggression towards healthcare staff in a UK general hospital: Variations among professions and departments. *Journal of Clinical Nursing* (13) pg 3-10.

Wells J and Bowers L (2002). How prevalent is violence towards nurses working in general hospitals in the UK? *Journal of Advanced Nursing* 39 (3) pg 230-240.

This page is intentionally blank.

12

Violence and aggression towards health care staff - II

BECKY MONAGHAN

Introduction

The BBC drama series 'Casualty' has been running for 19 years (www.bbc.co.uk/casualty). In that time the Accident and Emergency Department of fictitious Holby City Hospital has seen three staff deaths, two rapes and over 40 incidents involving violence and aggression towards staff, including stabbings shootings and other assaults. Is this a reflection of 'real life' in a National Health Service (NHS) Accident and Emergency (A&E) Department? Regardless of the truth of the matter, nevertheless television shows like Casualty have made violence and aggression towards health care staff feature more prominently in the public's collective consciousness.

The extent of violence and aggression within the NHS has been documented in 'A Safer Place to Work' (National Audit Office, 2003), which shows that in 2003 there were over 116,000 reported incidents of violence and aggression in the NHS as a whole compared to 65,000 in 1998. Indeed, according to the National Audit Office (NAO), violence and aggression against NHS staff is a serious issue that accounts for over 40% of all reported health and safety incidents. A previous report on health and safety in acute hospital trusts by the NAO (1996) highlighted concerns about the burden of incidents on the NHS, including violence and aggression, and the lack of information on the extent of incidents and their costs.

In reading the statistics alone, it is easy to assume that violence and aggression towards NHS staff is an increasing phenomenon, with the number of reported incidents in 2003 having risen to almost double that of 1998 (National Audit Office, 2003). However, if one examines the number of government initiatives since 1998 that have been aimed at increasing reporting and therefore reducing the number of incidents, it could be argued that the rise in numbers has been largely due to better reporting systems (Ferns and Chojnacka, 2005).

In 1998 the then secretary of state for health, Frank Dobson set national targets of a 20% reduction in violent incidents against NHS staff by 2001, and 30% by 2003 (Gourney, 2001). Further to this, health service managers were encouraged to prosecute the perpetrators of violence towards staff under the banner of 'Zero Tolerance' (Department of Health, 1999a; Department of Health, 1999b), the aims of which were two fold – "reinforcing to the public the message that violence towards staff working in the NHS was not acceptable, and also reassuring staff that violence and intimidation are unacceptable and will no longer be tolerated" (Beech and Leather, 2003). According to the National Audit Office (2003), the Zero Tolerance campaign increased the reporting of violent incidents by 30%, which Ferns and Chojnacka (2005) believe is a more realistic impression of clinical practice. However, the National Audit Office believe that incidents are still underreported in the NHS (National Audit Office, 2003). This supports a paper produced by the International Council of Nurses (1999), which estimated that only 20% of violent incidents were ever reported.

Ferns and Chojnacka (2005) produced a list of reasons why nurses in particular fail to report all violent incidents, and this list is reproduced as Figure 12.1. They based this list on evidence produced by Cembrowitz and Shepherd 1992, Kozlowska *et al* 1997, Beale *et al* 1999, Gournay 2001, Henry and Ginn 2002, and Lynch *et al* 2003.

Reasons for under-reporting violent or aggressive incidents

- The frequency and number of violent incidents is so great that the issue goes unreported – experiencing violence is routine
- Previous reporting of incidents has not led to change, so staff believe that reporting incidents is not worthwhile
- Reporting procedures is time-consuming
- Nurses fear they will be accused of negligence and inadequate performance
- Lack of agreement on definitions of violence
- Lack of awareness of the reporting system
- The belief that the incident was not serious enough to report
- The perception that nurses are hardened or desensitised to violence and perceive violence as 'part of the job'
- Excessive workloads
- Nurses wishing to avoid blame by colleagues or administrators
- Beliefs that the perpetrator was provoked by staff
- The practice of not reporting 'unintentional' violence, for example, that caused by confused or disorientated patients
- The nursing ethic of coping – there is evidence that nursing staff manage aggression passively, empathising with patients and hence avoid blaming them. Different grades of staff also respond to aggressive incidents differently

(Cembrowitz and Shepherd 1992, Kozlowska *et al* 1997, Beale *et al* 1999, Gournay 2001, Henry and Ginn 2002, Lynch *et al* 2003)

Figure 12.1. After Ferns and Chojnacka, 2005

According to Hollin (1992), the level of under reporting of violent incidents to NHS staff appears to coincide with the under reporting of violent incidents in general. However, further papers have been produced highlighting an increase in violence towards healthcare workers. In 2005, West Midlands Ambulance Service experienced a three per cent increase in the number of violent incidents towards staff and were concerned that their staff are 'now more susceptible to violent conduct and aggressive behaviour than ever before' (Parry, 2005).

Violence and threatening behaviour are "commonplace" in mental health and learning disabilities units, according to the Healthcare Commission (2005). The report also found that 40% of clinical staff and 80% of nursing staff as a whole have been victims of violent or threatening behaviour. Unison declared that staff face 'blows and abuse on the job' and called for a new law to protect health care workers, claiming that 'staff are punched, kicked, spat at and abused…..all for doing their job….' (Unison 2004).

Darymple (2002) stated that apart from the reported incidents of violence in hospitals, greater levels of incivility exist, where "loutish patients insist upon dropping litter in hospitals, having the TV on at full blast regardless of what other patients may want and talking noisily on mobile phones in busy wards or taking calls during consultations. They use emergency ambulances as a taxi service – one man calling for the ambulance 150 times in one year with impunity - and all but have sex with their visiting partners in multi-person wards. Doctors, nurses, and patients who object to such behaviour are subject to further verbal abuse, or worse." He believes that it is the toleration of these more minor incidents that leads to the increase in violence towards doctors and nurses. He also comments on the fact that hospitals have trained security guards and many have police officers stationed within them; and that managing violence and aggression is part of mandatory training, even going so far as to state "Henceforth a knowledge of karate will be as important to a doctor as a knowledge of pharmacology; and in the process, hospitals will have changed from being silent sanctuaries from the hurly-burly of life to being armed camps under permanent siege from the barbarians" (Darymple 2002).

The Department of Health's *Zero Tolerance Campaign* and associated *Managers' Guide* (Department of Health, 1999c) highlighted the importance of reporting incidents involving violence towards NHS staff. The campaign also reinforced the message that the Department of Health considers violence against NHS staff as an important issue. The campaign appears to have been successful in raising awareness amongst the public that violence against NHS staff will not be tolerated.

Munro (2002) in her paper 'Why do nurses neglect to report violent incidents?' believes that nursing is arguably the most dangerous job in the UK, but the rate of reporting is much less than the rate of actual incidents. Her research suggests that the severity of the incident is what determines whether nurses feel the incident should be reported. The findings of Nobel et al (1989), Pearson et al (1986) and Nelson et al (1997) all suggest that the reporting of incidents by nurses is inconsistent, particularly in psychiatry.

Whittington et al (1996) in their paper 'Violence to Staff in a General Hospital Setting' looked at the prevalence of violence and aggression in general health settings. They found that 21% of staff, responding to a questionnaire on their experience of violence, had been physically assaulted and 90% of these assaulted staff worked beyond the accident and emergency department, e.g. in medical wards. They also found that nurses were physically assaulted, threatened and verbally abused at higher rates than other professionals.

Wells and Bowers (2002) undertook a study into the prevalence of violence towards nurses, and found that nurses do face a high level of risk when compared with other professions. Duxbury and Whittington (2005) looked at the causes and management of violence and aggression towards nurses and found that the nurses surveyed felt that mental illness was the greatest cause of violent incidents.

The *Management of Health, Safety and Welfare Issues for the NHS* (NHS Employers, 2005), also known as the 'Blue Book', provides guidance to NHS employees regarding the issues of violence and aggression. The document highlights that one in seven of all reported injuries at work in NHS trusts are physical assaults by patients or visitors. Those who are particularly vulnerable to aggressive behaviour include nurses, ambulance staff, A&E staff and carers of psychologically disturbed patients. General Practitioners (GPs) and their staff are also documented as victims of assaults.

The Health Development Agency (2001) have explored the issues surrounding violence to GPs and their associated practice staff. Their guidance promotes a five step approach to risk assessment, and suggests that healthcare workers are at greater risk from violence and aggression than the general population.

The HSE (Health and Safety Executive, 2003) produced a sector information minute (SIM 07/2003/10) concerned with violence and aggression in healthcare. The minute notes the rise in violence and aggression in healthcare, stating that only one in five trusts met NHS targets for reducing incidents involving violence and aggression in 2001/02.

Mental health and violence and aggression

The relationship between mental health and violence and aggression has been extensively researched. Monahan (1992) found that, despite taking into account all the social and demographic factors, there remains a link between mental disorder and the propensity towards violent behaviour. Morral (2000) also agrees with this sentiment. In his book 'Madness and Murder' he believes that mental health professionals should acknowledge the danger of acutely mentally ill patients and stop offering bland reassurances to the public.

Gale et al (2002) looked at violence in Psychiatric units in New Zealand and found there was much variation between different units and the types of interventions they performed, concluding this needed further research.

Tucker (2002) found in his paper 'The enigma of violence: developing a therapeutic response' that, following a UKCC report, there was no doubt that violence in mental health settings is common, that there is a high level of under reporting and that it poses great risks to patients and staff. Nolan et al (1999) looked at violence in mental health care from the perspective of nurses and psychiatrists, and found that it was a serious problem that appeared to be increasing.

Violent crime

If the relationship highlighted by Hollins (1992) between violent incidents in the NHS and violent crime in general is accepted as true, then evidence that changes to reporting mechanisms in general violent crime statistics have had a significant impact on the numbers reported should be taken into account when examining NHS statistics.

Thorpe and Ruparel (2005) state that "The rate of victims' reporting of crimes has remained broadly stable since 1997, whereas the rate of recording of crimes by the police has been increasing, especially in the last three years, largely as a result of the national introduction of the National Crime Recording Standard (NCRS)." The NCRS is a standard recording system adopted throughout all police forces in April 2002. It has been suggested, however, that continuing auditing and improvements to the level of implementation of the standard have led to a continued increase in the numbers of recorded crime.

According to Duxbury (2002) aggression is not a 20th century phenomenon. But, according to the Department of Health (1999) and Rippon (2000), there has been an unprecedented rise in violent behaviour reported in recent years. Other opinions such as McRobbie and Thornton suggest that there is a level of 'moral panic' that has been fostered by the media, although violent crime statistics have stayed relatively stable over the years (McRobbie and Thornton 1995).

In a study undertaken by the Economic and Social Research Council (ESRC) relating to violence towards community based professionals such as clergymen and GPs, the findings indicated that a generational difference in attitude was an important part of the perception of aggression. According to the study, "the perceived increase in violence is often attributed to their regrettable loss of status compared with the `golden age' of the 1950s and 1960s, when professionals were treated more deferentially." Therefore the feeling that 'it was never like this in my young days' is quite prominent (ESRC, 1996).

Alcohol and Violence

The relationship between alcohol and violence is a well documented one, although differing opinions exist as to whether alcohol increases the level of violence in society. Parker (2002) found that "There are complex but strong statistical relationships between alcohol consumption and crimes of violence in most western countries.'"

The American National Institute on Alcohol abuse and Alcoholism (NIAAA) state that "Alcohol may encourage aggression or violence by disrupting normal brain functions." This suggests that disinhibition takes over and brain mechanisms that normally restrain impulsive behaviours, such as violence, are weakened by alcohol (NIAAA, 1997). However, a paper from the Social Issues Research Centre (SIRC) in Oxford finds that 'there is no direct causal link between alcohol and violence. The probability of aggression is increased when the effects of alcohol-induced cognitive impairment are amplified or exacerbated by both the characteristics of the immediate situation and cultural expectations that drinking causes aggression. Where the immediate social context is non-aggressive and where cultural beliefs and norms inhibit aggression, drinkers are highly unlikely to become aggressive." (Social Issues Research Centre, 2005).

In countries where alcohol is more integrated into life, such as Spain or Italy, binge drinking and associated antisocial behaviours are rare, and therefore the relationship between alcohol and violence is questionable. However, it is suggested that alcohol and other sociological influences are associated with violence in the UK - usually in the form of binge drinking (MCM Research 2004).

Bellis et al (2005) looked at the effects of nightlife on the health of individuals, finding that alcohol had a significant effect on behaviour; and in a previous study Bellis et al (2004) found that "one fifth of violent incidents occur in or around pubs and clubs, with 80% of those involving alcohol."

Therefore, it could be concluded that alcohol does seem to have an effect on violence. Whether this is due to biochemistry, brain function or complex sociology remains open to debate.

Violence and Mental Illness

Anderson (2003) believes that throughout the past decade there has been an increase in media attention on issues relating to community care and the discharge of people from institutions. He further believes that newspapers in the UK have a great deal of influence when reporting violent incidents involving patients who have mental health problems.

Laurance (2003) comments that until a high profile homicide by a patient with schizophrenia in 1992, the UK's main concern related to mental health care was the well-being of patients discharged into the community following closure of long term mental institutions. After the homicide, concern shifted to public protection. However, Laurance points out that although an estimated 600,000 people in England have a diagnosed severe and enduring mental health problem such as schizophrenia, less than one per cent of these require intensive care as a result of being a risk to others. However, the rare cases of homicide that have taken place have led to increased fear amongst the public and a stigmatisation of people with mental illness. Laurence found that mental health and violence were not linked unless there was a combination of "dual diagnosis", which is alcohol or drug dependence and mental illness.

In contrast to these sentiments, Morrel (2000) states that mental health professionals should "acknowledge the danger of acutely mentally ill patients and stop offering bland reassurances to the public." Mental health charities such as Rethink and Mind have campaigned to try and reduce the stigma that exists around mental health and violence. Rethink joined forces with United International Pictures to "get the message out to the general public that schizophrenia is not the end of the world, is not about violence and that illness does not define who a person is" (Rethink, 2005). This followed the success of the film 'A Beautiful Mind' which tells the true story of John Forbes Nash who was diagnosed with schizophrenia at the age of 30, but went on to win the Nobel Prize for Economics in 1993.

Paterson et al (2004) found that a changing population has had an impact on the change of opinion from a previously universal consensus that mental illness and violence are linked to a newer school of thought that sometimes there is a link, although there are other contributing factors.

European and Global perspective

In previous sections I have already commented on how various authors have found the links between violence and culture in relation to alcohol consumption to be interconnected.

The World Health Organisation (WHO) reports that violence costs some countries more than four per cent of their annual gross domestic product (WHO, 2002). The WHO report found that 1.6 million people die as a result of violence each year with millions of others injured physically or psychologically. Colombia and El Salvador spend 4.3 per cent of their GDP on health costs related to violence, while Brazil spends 1.9 per cent and Peru 1.5 per cent, with industrialized countries also facing high economic costs. In Australia, for example, workplace violence costs US$837 million to the economy each year and $5,582 to employers for every victim. In one province of South Africa, Western Cape, homicides alone cost US$30 million each year.

Ferns (2005) found that nurses were equally susceptible to violent attacks internationally, but methods of assault varied, with, for example, more weapons being used in North America than the UK. Hegney et al (2003) in a study performed in Australia, found that the level of violence experienced by nurses was dependent on the specialty they worked in; a similar conclusion was arrived at in the UK by Whittington et al (1996).

Mental health links to violence in Scandinavia were discussed by Matthias and Angermeyer (2000), who found that there is 'a moderate but significant association between schizophrenia and violence'.

Comparison of Experience of Violence and Aggression within Mental Health

For this exercise, I undertook qualitative interviews with two mental health professionals of different age groups. The aim of the exercise was to try and establish how violence and aggression affected their working lives and to compare and contrast their experiences.

A common questioning spine was used, which consisted of four prompt words; experiences, comparisons, perceptions and support. Encouragement to discuss the four topic areas stated above was provided in the form of open questions. However, an attempt was made to avoid direct questioning, preferring instead to allow the subjects the opportunity of free dialogue and expression without the need to rely on questions and answers.

The first subject was female and in her late thirties, having trained as both a general and mental health nurse in the 1980's. The second was an older male who trained as a mental health nurse very early in his teenage years, then later undertook general nurse training.

Experiences

Subject 1 trained and worked predominantly in the NHS. She stated that the services she worked in were contemporary in their thinking and she was still involved in direct patient care and was 'client-facing'. Her experiences of violence and aggression were numerous, suggesting that was the nature of the client group. On separate occasions she had received injuries such as a broken nose and displaced teeth. She had experienced direct threats to her safety, which she found more frightening than physical assaults. She had been imprisoned in her office by a client, who smashed windows by throwing missiles such as house bricks towards them, whilst she tried to protect herself by hiding under a desk.

Subject 2 worked as a cadet nurse from the age of fourteen in a traditional NHS mental hospital. He experienced many instances of violence and aggression, again relating to the client group. When Subject 2 was seventeen, a patient caused a crush injury to his feet, leaving him with a permanent disfigurement and impediment. Aged eighteen, Subject 2 was hospitalised following a serious assault by a patient that left him temporarily paralysed, and side effects remain.

Perceptions

Subject 1 stated that when working in the stressful environment where she experienced many instances of violence and aggression, her perceptions of what was normal were altered. It became normal to deal with high levels of expressed emotion, some of this manifesting itself physically. She talked about experiencing a kind of 'siege mentality', with the team she worked with protecting each other and just getting on with the work that needed to be done. She also stated that she felt frightened on many occasions, but she felt more intimidated by people she had contact with that weren't clients, such as relatives.

Subject 1 also stated that fear of the unknown was quite dominant in her feelings, and she felt more fear from threats than physical assault. She said that those clients who were ill caused her less anxiety than those with personality disorders.

Subject 1 also made comments on the nature of the culture of the departments she worked within, finding there to be a 'bullish male culture', where staff provoked clients by confrontational behaviour and some of the clinical practices, she felt, were antagonistic. As part of the interview, we also discussed the types of roles people adopt when dealing with a violent or aggressive situation. These discussions led me to develop the model discussed later in this chapter.

Subject 2 also stated that he felt fearful on many occasions, commenting that it was sensible to do so as some of the patients he cared for were 'killers'. He described the environment he worked in with stronger language than Subject 1, choosing words such as 'brutal' and 'violent'. He also described a 'military' culture where it was accepted that individuals would 'go into harm's way' as a daily part of their job. He said that some staff, including himself, were occasionally antagonistic and arrogant when dealing with patients. He described an environment that was confrontational, and said that staff were influenced by this and became either 'angels or bastards' towards the clients.

Subject 2 stated that many treatments and certain illnesses caused confrontation between staff and clients, and some staff handled this badly, tending to be overly controlling. He said that this made him modify his actions to reduce this situation and he gave these clients careful handling so as not to provoke a violent reaction.

Comparisons

Subject 1 felt that her experiences were not unlike those of her peers, although she did suggest that certain people assumed roles, such as that of a hero or victim, and therefore this caused them to receive more incidents of violence. Subject 2 agreed with this hypothesis, suggesting again that some people provoked clients into reaction. He stated that violence was never intellectualised about in the workplace and was not seen as anything out of the ordinary. He felt that psychiatry was still a very young science and there wasn't a school of thought about violence or aggressive behaviour.

<u>Support</u>

Subject 1 felt that she was very well supported by her peers and her management structure. There existed an established method of clinical supervision that allowed for debriefing after serious incidents.
She felt that individuals also developed negative methods for coping with the stress of the aggressive environment, such as smoking and drinking, but found other coping mechanisms such as humour much healthier. There was a high level of sickness absence due to stress related conditions in the area where Subject 1 worked.

Subject 2 described a very different scenario, of a very traditional hospital with rituals and procedures as well as an expectation that individuals would cope. Staff were recruited on their size and ability to be 'tough', and the wards were very masculine and treatments would now be seen as quite brutal. He described a situation where individuals who had been assaulted were treated like heroes but stated that he would hide in a side room and cry after particularly bad incidents. There was no sickness absence due to stress; this was not an accepted action or part of the mindset of the staff he worked with.

Violence and Aggression Model

Following a review of the relevant literature around behaviour models and discussions held during the qualitative interviews above, I found that there does not appear to be a model that describes the roles undertaken in violent and aggressive situations.

Karpman (1968) produced a model for social interactions known as the drama triangle (Figure 12.2). Forest (1996) gives an interpretation of this model, suggesting that individuals undertake each of these roles within a social interaction setting. She suggests that victim is at the bottom as all roles are different facets of victim, with all parts receiving a negative action within the social interaction.

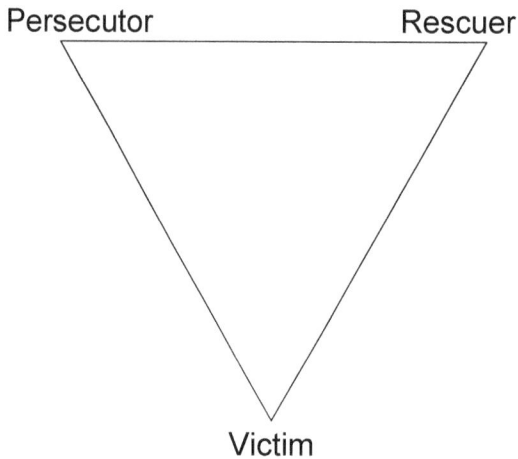

Figure 12.2 - Drama Triangle

I believe that the drama triangle can be adapted to better represent a violent and aggressive situation. If one considers the simplistic model of cognitive dissonance known as Betari's Box (Figure 12.3), where it is suggested that an individual's attitude affects their behaviour, and therefore the attitude of another person and subsequently their behaviour, a further dimension of 'antagonist' can be added to the triangle (Figure 12.4). This would also concur with the experiences discussed in the qualitative interviews above that some individuals antagonised patients and caused a violent response.

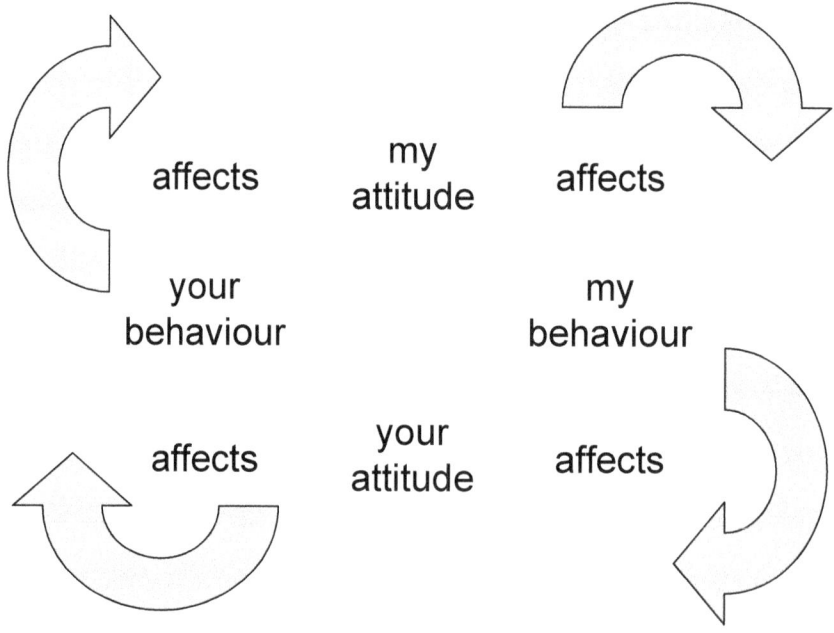

Figure 12.3 – Betari's Box

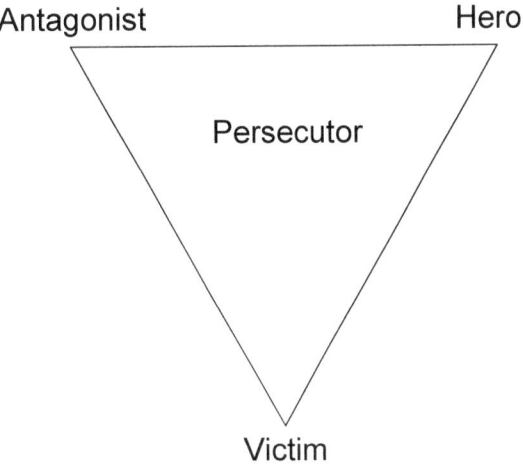

Figure 12.4 – Adapted Drama Triangle

In this adapted model, a further dimension of antagonist is added and the role of persecutor is centralised to suggest that role is taken by another person. The theory behind this model is that the roles of antagonist, hero and victim are the three facets of the victim role, whilst the persecutor is the individual performing the violent or aggressive act.

The hero is a person who adopts the role of leader; perhaps they have more experience of dealing with violent situations, maybe they are male when other staff members are female and feel an obligation to protect those perceived less able than themselves. In this way, the hero becomes a victim, by inviting the persecutor to select them as the person to receive the violent act.

The victim is a person who is perhaps quite diminutive, vulnerable and an easy target, therefore the persecutor selects them to be the receiver of the violent act. The antagonist is as its name suggests a provoker of response by confrontation; and therefore becomes the victim when the persecutor reacts and selects them as the receiver of the violent act.

Conclusions

There is evidence to suggest that violence is becoming more commonplace in the NHS and society in general, although repeated government campaigns to raise awareness of violence towards NHS staff have led to a more proactive workforce in relation to reporting.

There are factors in society that affect the level of violence that cannot be ignored, such as alcohol and mental illness, although the causal links between both these factors and violence are very much open for debate.

The qualitative interviews described in this chapter give an insight into the changing field of mental health nursing and its attitude to violence and aggression; although the prevalence of violence appears to be the same now as it has been over the years, the response to it is very different.

References

Anderson, M (2003). One flew over the psychiatric unit: Mental illness and the media. *Journal of Psychiatric and Mental Health Nursing* 10: 297-306.

Beale D, Leather P, Cox T, Fletcher B (1999). Managing violence and aggression towards NHS staff working in the community. *Nursing Times Research* 4, 2, 87-99.

Beech B and Leather P (2003). Evaluating a management of aggression unit for student nurses. *Journal of Advanced Nursing* 44, 6,603-612.

Bellis M A, Hughes K, Ashton J R (2004). The promiscuous ten percent? *Journal of Epidemiology and Community Health* 58, 11, 889-890.

Bellis M A, Hughes K, McVeigh J, Thomson R, Luke C (2005). Effects of nightlife activity on health. *Nursing Standard* 19, 30, 63-71.

Paterson B, Claughan P and McComish S (2004). New evidence or changing population? Reviewing the evidence of a link between mental illness and violence. *International Journal of Mental Health Nursing* 2004 13:1 39.

Gale C, Pellett O, Coverdale J, Paton-Simpson G (2002). Management of violence in the workplace: a New Zealand survey *Acta Psychiatrica Scandinavica* 2002 106: s412 p41.

Canning U P et al (1999). Substance misuse in acute general medical admissions. *Quarterly Journal of Medicine* 92: 6, 319-326

Cembrowitz S P, Shepherd J P(1992). Violence in the accident and emergency department. *Medicine, Science, and the Law* 32, 2,118-122.

Darymple T (2002). *Research Report 36 Violence, Disorder And Incivility In British Hospitals The Case For Zero Tolerance.* The Social Affairs Unit. Great Britain.

Department of Health (1997). *Health and Safety in NHS Acute Hospital Trusts in England.* Health Service Circular 1996/82. www.dh.gov.uk

Department of Health (1999a). *NHS zero tolerance zone campaign - Press Statement.* 14 October 1999. www.doh.gov.uk

Department of Health (1999b). *Campaign to stop violence against staff working in the NHS: NHS zero tolerance zone.* Health Service Circular 1999/226. www.dh.gov.uk

Department of Health (1999c). *NHS Zero Tolerance Zone: We Don't Have To Take This.* Resource pack. London: Stationery Office

Duxbury J (2003). *National Directives on managing violent patients; A critique.* Nursing Times Vol 99 No 06.

Duxbury J and Whittington R (2005). Causes and management of patient aggression and violence: staff and patient perspectives. *Journal of Advanced Nursing* 2005 50:5 469.

Economic and Social Research Council (1996). *Violence against Professionals in the Community.* Violence Research Programme London.

Felson R B and Pare P (2005). The Reporting of Domestic Violence and Sexual Assault by Nonstrangers to the Police. *Journal of Marriage and Family* 2005 67:3 597.

Ferns T, Chojnacka I (2005). Reporting incidents of violence and aggression towards NHS staff. *Nursing Standard* 19, 38, 51-56.

Ferns T (2005). Violence in the accident and emergency department - An international perspective. *Accid Emerg Nurs.* 2005 May 28.

Forest L (1996). The Drama Triangle: The Three Faces of Victim. http://lynneforrest.com/html/the_faces_of_victim.html#victim

Gordis E (1997). *Commentary.* National Institute on Alcohol Abuse and Alcoholism (NIAAA). USA. http://pubs.niaaa.nih.gov/publications/aa38.htm

Gournay K (2001). *The Recognition, Prevention and Therapeutic Management of Violence in Mental Health Care. A Consultation Document.* Department of Health Services Research, Institute of Psychiatry and South London and Maudsley NHS Trust, London.

Harris R (2004). Media representation of people with mental health problems. *Nursing Times* Vol. 100, No. 34, August 2004.

Health and Safety Executive (2003). Sector Information Minute SIM 07/2003/10. www.hse.gov.uk

Healthcare Commission (2005). *National Audit of Violence 2003-5*. Healthcare Commission. London. www.healthcarecommission.org.uk

Health Development Agency (2001). *Violence and Aggression in General Practice*. National Institute for Health and Clinical Excellence. London. www.nice.org.uk/page.aspx?o=502103

Health Education Authority (1998). Perceptions of alcohol related attendances in A&E departments in England: a national survey. *Alcohol and Alcoholism* 33: 4, 354-361.

Henry J, Ginn G O (2002). Violence prevention in healthcare organisations within a total quality management framework. *Journal of Nursing Administration*. 32, 9,479-486.

Hollin C R (1992). *Criminal Behaviour*. Falmer Press Basingstoke.

International Council of Nurses (1999). *Increasing Violence in the Workplace is a Threat to Nursing and the Delivery of Health Care* (Press release). March 8. Geneva. www.icn.ch/PRviolence_99.htm

Wells J and Bowers L, (2002). How prevalent is violence towards nurses working in general hospitals in the UK? *Journal of Advanced Nursing* 2002 39:3 230.

Leonard KE (2002). Alcohol's role in domestic violence: a contributing cause or an excuse? *Acta Psychiatrica Scandinavica* 2002 106:s412 9.

Karpman S. (1968). Fairy Tales and Script Drama Analysis. *Transactional Analysis Bulletin* vol. 7, no. 26, pp. 39-43.

Kozlowska K, Nunn K K, Cousens P (1997) Adverse experiences in psychiatric training: Part 2. *Australian and New Zealand Journal of Psychiatry* 31, 5,641-652.

Laurance J (2003). *Pure Madness. How Fear Drives the Mental Health System.* London: Routledge.

Lynch J, Appelboam R, McQuillan P J (2003). Survey of abuse and violence by patients and relatives towards intensive care staff. *Anaesthesia* 58, 9, 893-899.

Marsh P (1996). Football Violence and Alcohol. Social Issues Research Centre. Oxford.

Matthias, C (2000). Angermeyer Schizophrenia and violence *Acta Psychiatrica Scandinavica* 2000 102:s407 63.

MCM Research (2004). Report of research and consultation conducted by MCM Research Ltd for Wine Intelligence. September 2004. www.sirc.org/publik/binge_drinking.pdf

McRobbie A, Thornton T (1995). Rethinking moral panic for multi-medial social worlds. *British Journal of Sociology* 46:4, 559-574.

Monahan J (1992). Mental disorder and violent behaviour: perceptions and evidence. *American Psychologist* 47, 3, 511-521.

Morral P (2000). *Madness and Murder*. Oxford University Press.

Munro V (2002). Why do nurses neglect to report violent incidents? *Nursing Times* Vol 98, No 17, April 2002.

Naevdal F (2005). Acceptance of physical violence (APV) among adolescents in a Norwegian normal sample; statistical description of the assessment. *Journal of Adolescence* 28 (2005) 425–431.

National Audit Office (1996). *Health and Safety in NHS Acute Hospital Trusts in England.* www.nao.gov.uk

National Audit Office (2003). *A Safer Place to Work: Protecting NHS Hospital and Ambulance Staff from Violence and Aggression.* www.nao.gov.uk

National Institute on Alcohol Abuse and Alcoholism (1997). *Alcohol Violence and Aggression.* Alcohol Alert No. 38, October 1997.
http://pubs.niaaa.nih.gov/publications/aa38.htm

Nelson D et al (1997). A descriptive study of violent incidents amongst in-patients in a psychiatric hospital. *Health Bulletin* 55, 6.

NHS Employers (2005). *The management of health, safety and welfare issues for NHS staff.*
www.pasa.doh.gov.uk/medicalconsumables/sharps/blue_book_complete%5B1%5D.pdf

Nicholas S, Povey D, Walker A and Kershaw K (2005). *Crime in England and Wales 2004/2005.* Home office Statistical Bulletin. London.
www.homeoffice.gov.uk/rds/pdfs05/hosb1105.pdf

Noble P and Rodger S (1989). Violence by psychiatric in-patients. *British Journal of Psychiatry* 155: 384-390. http://bjp.rcpsych.org/cgi/content/abstract/155/3/384

Parker R N (2002). *Alcohol and violence- what's the connection?* University of California. www.aphru.ac.nz/hot/violence.htm

Parry (2005). *Violence and Aggression on the Increase.* West Midlands Ambulance Service Press Release. www.wmas.nhs.uk/Home/tabid/89/Default.aspx

Pearson M, Wilmot E and Padi M (1986). A study of violent behaviour among patients in a psychiatric hospital. *British Journal of Psychiatry* 149, 232-235. http://bjp.rcpsych.org/cgi/content/abstract/149/2/232

Nolan P, Dallender J, Soares J, Thomsen S, Arnetz B (1999). Violence in mental health care: the experiences of mental health nurses and psychiatrists *Journal of Advanced Nursing* 1999 30:4 934.

Proctor D (2003). Alcohol Related problems in general hospital. *Nursing Times* Vol 99, No 09.

Rassool G H, Winnington J (2003). Adolescents and alcohol misuse. *Nursing Standard.* 17, 30, 46.

Rethink (2005). *Respect your beautiful mind.* www.rethink.org

Rippon T J (2000). Aggression and violence in health care professions. *Journal of Advanced Nursing* 31:2, 452-460.

Social Issues Research Centre (2005). *Alcohol and Violence.* Oxford. www.sirc.org

Stanko B (1996). *Violence against professionals in the community.* Economic and Social Research Council. London. www.esrc.ac.uk/ESRCInfoCentre/research/research programmes/violence.aspx?ComponentId=9259&SourcePageId=9102

Thorpe K and Ruparel C (2005). *Reporting and recording crime.* Chapter in *Crime in England and Wales 2004/2005.* Home office Statistical Bulletin. London. www.homeoffice.gov.uk/rds/pdfs05/hosb1105.pdf

Tucker R (2002). The enigma of violence developing a therapeutic response *Mental Health Practice* February 2002 Vol 5 No 5.

Unison (2004). *Union calls for new laws to protect health workers.*
www.unison.org.uk/news/news_view.asp?did=1820

Whittington R, Shuttleworth S and Hill L (1996). Violence to staff in a general hospital setting. *Journal of Advanced Nursing* 24: 326-333.

World Health Organisation (2002). *World Report on Violence and Health.*
www.who.int

Hegney D, Plank A and Parker V (2003). Workplace violence in nursing in Queensland, Australia: A self-reported study. *International Journal of Nursing Practice* 2003 9:4 261.

13

Fire safety and the training of staff in fire prevention and management in healthcare premises

JAYNE HARTLEY

Introduction

This chapter explores whether the training provided for healthcare staff in the prevention and management of fire requires restructuring, and whether the current format for fire prevention and management training is research based and educationally sound. To underpin this exploration, the current arrangements for fire training in the National Health Service (NHS) are examined and a review of current fire training methods in five NHS Trusts is presented.

The incidence of fires occurring in the NHS is considered, including fire alarm reported incidents within the author's place of employment, the Christie Hospital NHS Trust. This will help highlight the importance of effective and meaningful fire prevention and management training.

I then review the effectiveness of fire training within the Christie Hospital through examining attendee evaluations of current mandatory fire training and conducting a questionnaire survey of senior managers in the Trust. The survey assessed managers' perceptions of the effectiveness and relevance of fire training within their own departments. In addition, it identified whether managers have assessed the risk of fire in their areas of responsibility and added this risk and subsequent action plan to their departmental risk register. The aim is to determine if fire safety is part of a culture of safety at the hospital, or is isolated to a single annual training session which has little impact on an individual's working practice. The survey also seeks to determine managers' thoughts as to what would make fire training more effective and how the content and delivery of the training sessions could be improved. Adopting the ideas of staff working in departments may help to increase the validity of fire training sessions and enhance interest in fire prevention and management.

The chapter concludes by considering how the current provision for fire prevention and management training for healthcare staff.

Fire training in the National Health Service

Until 1990, NHS premises fell under the scope of Crown Immunity, which meant they did not need to comply with the 'letter of the law' relating to fire safety (Roberts, 2003). However, following the NHS and Community Care Act 1990, Crown Immunity was fully removed in April 1991.

There appears to be some confusion in the literature as to the significance of the impact of the NHS and Community Care Act 1990 in this area. NHS Estates (1994a) highlight that NHS organisations had already lost crown immunity in 1987, when they became subject to the full provisions of the Health and Safety at Work Act 1974. The relevance of crown immunity is also questioned by Roberts (2003), who feels that the NHS has maintained a good record in respect of fire safety for many years, as shown by relatively low number of fire deaths and injuries.

Nonetheless, in 1990, all NHS organisations, their staff and fire prevention advisers were required to ensure compliance with the mandatory fire safety requirements of 'Firecode' (Roberts, 2002, p.182; Roberts, 2003). Firecode is a suite of documents, formerly published by NHS Estates with the intention of providing a systematic approach to reduce the potential for fire in health service premises [**Editor's note**: With the demise of NHS Estates, information on Firecode etc. can now be obtained via the Department of Health website at www.dh.gov.uk]. The code sets standards for the layout, design, construction and fire safety management of hospitals and other healthcare premises. Firecode is underpinned by a 'policy and principles' document (NHS Estates 1994a), and includes a number of Health Technical Memoranda (HTMs) and Fire Practice Notices (FPNs) which consider policy, technical guidance and specialist aspects of fire precautions (NHS Estates, 2004). The full set of codes is listed in table 13.1.

Firecode (NHS Estates, 1994a) stresses the requirement for a fire safety strategy that is based primarily on the avoidance of fire. It also states that the strategy must be supported by a procedure for staff training and re-training, but does not elaborate further on what this training should entail.

	Firecode - policy and principles (March 1994)
	Fire risk assessment in Nucleus hospitals (March 1997)
HTM 81	Fire precautions in new hospitals (April 1996)
HTM 82	Alarm and detection systems (September 1996)
HTM 83	Fire safety in healthcare premises - general fire precautions (April 1994)
HTM 85	Fire precautions in existing hospitals (April 1994)
HTM 86	Fire risk assessment in hospitals (April 1994)
HTM 87	Textiles and furniture (April 1999)
HTM 88	Fire precautions in housing providing NHS - supported living in the community (June 2001)
FPN3	Escape bed-lifts
FPN4	Hospital main kitchens
FPN5	Commercial enterprises on hospital premises
FPN6	Arson prevention and control in NHS premises
FPN7	Fire precautions in patient hotels
FPN10	Laboratories on hospital premises
FPN 11	Reducing unwanted fire signals in healthcare premises

Table 13.1: Current Firecode Suite of Documents
(Source NHS Estates, 2004)

Of the seven HTMs and seven FPNs that currently contribute to Firecode, the issue of training is referred to in only three HTMs and three FPNs. The largest section on training is contained within HTM 83 *Fire safety in healthcare premises - general fire precautions* (NHS Estates, 1994b). This covers the need to ensure staff are aware of the principles of fire prevention, fire hazards, fire fighting equipment and the action to be taken in the event of a fire. There is also some discussion within HTM 83 about who should carry out fire training, the need to maintain attendance records and to evaluate training sessions, consideration of the practicability of carrying out fire drills, the provision of a suitable induction for all new starters and the requirement for all staff to attend an hour's fire safety training session each year. This training requirement is mandatory and concern has been raised over the prescriptive nature of this aspect of the guidance which may not be achievable or necessary (Roberts, 2003).

There are also a number of Acts and Regulations that have a bearing on fire safety in NHS healthcare premises. The most relevant are identified in table 13.2 (Roberts, 2002 p.181)

> The Fire Services Act 1947
> The Health and Safety at Work etc. Act 1974
> The Fire Precautions Act 1971, as amended by the Fire Safety and Safety of Places Sports Act 1987
> The Building Act 1984
> The Registered Home Act 1984
> The Housing Act as amended by the Local Government and Housing Act 1989
> Fire Precautions (Workplace) Regulations 1997

Table 13.2 – Acts and regulations related to fire safety in NHS premises (Source – Roberts, 2002 p.181)

The *Health and Safety at Work etc. Act 1974* places a general responsibility on employers for the protection of the health, safety and welfare of their employees, which includes the provision of safe means of access and egress (NHS Estates 1994a). However, the frequency or type of fire safety training is not referred to in this legislation, apart from specifying that staff should be trained in the use of fire extinguishers – which may not necessarily be a priority.

The *Fire Precautions (Workplace) Regulations 1997*, and amendments, set out regulations for workplaces that focus primarily on the need to undertake risk assessments (and this links closely with HTM 86 - *Fire risk assessment in hospitals* - NHS Estates, 1994c). However, these regulations make only passing reference to the need to ensure 'adequate' training for 'employees'. No guidance is given on what might constitute 'adequate' training.

A further influencing factor on fire training in health care was the former controls assurance standard for fire safety (Department of Health – now withdrawn). Controls assurance was a process aimed at providing evidence that NHS organisations were doing their reasonable best to meet their objectives and to protect patients, staff and the public against risks of all kinds (Health Care Standards Unit, 2004). Criterion 16 of the former controls assurance fire safety standard required all staff to receive a level of fire safety training appropriate to their individual responsibilities in the event of a fire. Training was to take place on induction and at least once more each year, with reference to HTM 83 for further guidance. However, this element of persuasion to provide training has diminished since controls assurance ceased to exist as a Department of Health policy initiative in 2004 (Health Care Standards Unit, 2004).

An outline of fire safety training in five NHS Trusts

This section briefly considers the fire safety training provided in a sample of five NHS Trusts: The Christie Hospital NHS Trust, the Royal Devon and Exeter NHS Foundation Trust, the Norfolk and Norwich University Hospital NHS Trust, Ashford and St Peter's Hospitals NHS Trust, and Newcastle Hospitals NHS Trust.

<u>1. Christie Hospital NHS Trust</u>

The Christie Hospital NHS Trust fire safety policy (as at October 2004) states the following in respect of fire safety training:

1. Training should be provided annually for all members of staff.
2. Training will, as far as practicable, be related to the particular discipline of the staff to be trained – for example, nursing staff will receive instructions and training in methods of evacuation.
3. Evacuation drills will be carried out at least twice per year in a selected area of the hospital, including patient care areas. The aim is to test both day and night staff.

According to the trust fire officer, fire lectures are held in the lecture theatre every fortnight. These sessions are scheduled to last for one hour and cover legislation, causes of fire, fire prevention, what to do in the event of a fire and different fire fighting equipment. The rationale behind the content of the course is based on legislative requirements and time available. Despite the intention in the fire safety policy to adapt the training to staff groups, this has not yet happened. In addition, evacuation drills have not been undertaken at the Christie hospital for the past two years due to staffing issues. Data are not currently available to indicate the percentage of staff that have received their annual mandatory fire training.

The trust fire officer used to be employed as the head porter and at the time of writing had been in post for 12 months. He does not have a teaching qualification.

<u>2. Royal Devon and Exeter NHS Foundation Trust</u>

The fire precautions policy issued by Royal Devon and Exeter NHS Foundation Trust (2004a) refers simply to the fact that every staff member must participate in fire training as directed by Firecode. There is no other reference to fire training. However, the annual fire report (Royal Devon and Exeter NHS

Foundation Trust, 2004b) expands on the variety of fire training that takes place. This includes formal training sessions, departmental sessions, induction and evacuation drills (of which there were four in 2002/2003). Over the past six years, attendance at the mandatory fire training sessions has averaged 73 percent.

According to the trust fire advisor, fire training follows the requirements of Firecode. The fire advisor used to work for the fire brigade and has been in post since December 1995. He does not have a formal teaching qualification.

3. Norfolk and Norwich University Hospital NHS Trust

Norfolk and Norwich University Hospital NHS Trust issued their fire safety policy in March 2004 and it refers to training in section 3 (Norfolk and Norwich University Hospital NHS Trust, 2004). The policy takes extracts from HTM 83 to explain what fire training is required, viz:

- understanding the causes of fire,
- fire hazards,
- fire prevention,
- action to be taken in the event of a fire and
- evacuation procedures.

This training is provided on induction and on an annual basis. Fire drills are also planned and specialist training which is carried out on an 'as and when required' basis, is provided for those staff with specific responsibilities in the event of a fire. The fire safety manager advised that fire drills have not been carried out over the past year due to staffing issues and that mandatory attendance at fire lectures for all staff is just over seventy percent. The training provided is formal and follows the guidance provided in HTM 83.

The fire safety manager used to be employed as a fire officer in the fire service. He has been in his current post for 13 years and has extensive experience and knowledge of fire prevention and management but does not have a formal teaching qualification.

4. Ashford and St Peter's Hospitals NHS Trust

The fire policy and procedural notes from Ashford and St Peter's Hospitals NHS Trust (2003) refers to training in section 9. There is a requirement that all staff must attend annual fire training, and that new staff receive fire training as part of their induction. Training content varies depending on whether a staff member is ward based or not. Ward based staff also receive a one day fire evacuation training day. It is also intended that fire drills should be held at least twice a year in each of the two hospitals.

The hospital fire prevention officer has based the content of the training days on legislation and feedback. The intention to run different sessions for clinical and non clinical staff arose as a result of staff feedback and this seems more beneficial. Fire drills are held as planned, which also helps to highlight areas where improvements can be made to overall fire safety procedures. However, it remains difficult to ensure that all staff attend a fire training session annually as required.

The hospital fire prevention officer in this trust used to be employed as a health and safety advisor, but has always had a keen interest in fire prevention. He has been in post for 27 years and has developed close links with the fire service during that time. He has no formal teaching qualification.

5. Newcastle Hospitals NHS Trust

An annual fire report is produced, which includes a section on fire training. The report highlights that fire training is mandatory and that each member of staff must receive fire prevention training at least once each year. It is also expected that all staff should take part in a fire evacuation drill, or a walk/talk through explanation and are given the opportunity to practise using the commonest fire extinguishers.

The report further highlights that in house fire briefs are carried out which involve training in an employee's own workplace. These are reported to be popular and meet mandatory training requirements. There are also more formal training sessions held monthly throughout the Trust, which take place during the day, at night and at weekends. In addition, a number of fire evacuation exercises have taken place within the Trust.

However, despite the mounting of many training sessions, at the time of writing only 63 percent of staff at the Newcastle Hospitals NHS Trust received their mandatory training. The fire advisor has attempted to address this issue with departmental managers and is planning to change the timing of the fire training sessions from one day per month to four days in succession every four months. However, the content of the training is planned to remain the same and is based on legislation, not research.

The fire advisor has been in post for seven years and was previously employed in the fire service. He has teaching experience both from his current and previous roles but has never undertaken a teaching course.

Discussion

The information collected from the above trusts provides a small but significant insight into the arrangements made in practice to deliver fire training. It can be seen that the basis for fire training is legislation not research, and that there is no standardised approach to the delivery of fire training. It can also be seen that none of the above trusts can meet the mandatory target of all staff members attending fire training annually. The staff highlighted above who are responsible for the provision of staff training do not have an educational qualification which contributes to the debate that current fire training sessions are not educationally sound. However, there is little doubt that four of the five fire trainers have extensive knowledge about, and teaching experience of, fire prevention and management.

Incidence of fires within the healthcare setting

The Office of the Deputy Prime Minister produces national statistics relating to fires in the United Kingdom (UK). In 2003, local authority fire and rescue services attended nearly 1.1 million fires or false alarms, 10 per cent more than in 2002. Within this figure, fires increased by 20 per cent to 621,000, while false alarms fell by one per cent to 473,000 (Office of the Deputy Prime Minister, 2005).

Of the 621,000 fires that occurred, approximately 105,000 took place in buildings such as homes, hotels, schools, offices and hospitals. In 2003, there were approximately 2,400 fires which occurred in hospitals. However, as shown in Figures 13.1 and 13.2, these figures have not altered significantly over the past 10 years.

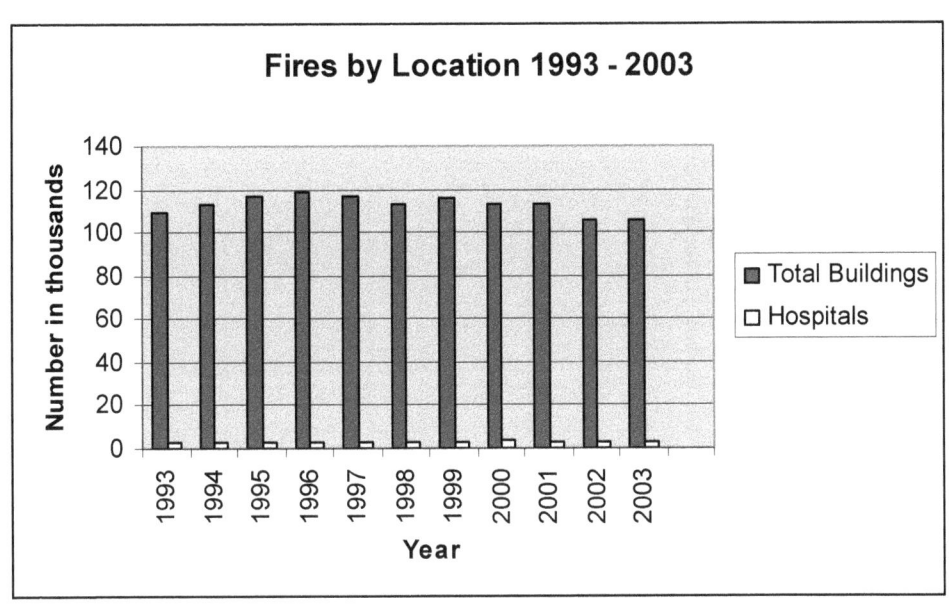

Figure 13.1 – Fires by location 1993 – 2003
(Source: Office of the Deputy Prime Minister 2005)

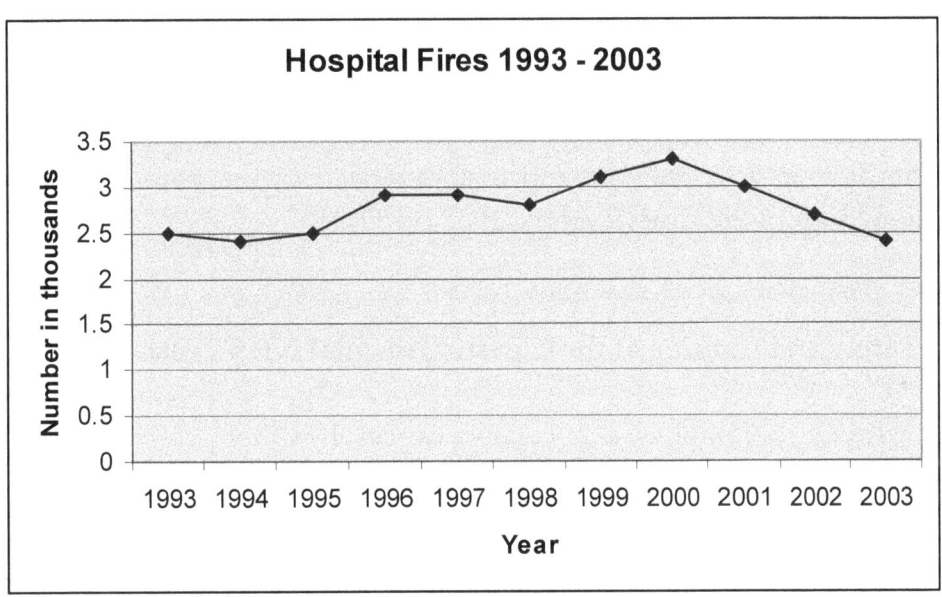

Figure 13.2 – Hospital fires 1993 – 2003
(Source: Office of the Deputy Prime Minister, 2005)

Unfortunately the information provided in these annual reports does not provide a breakdown of deaths and non fatal injuries in hospitals, apart from a review of deaths and non fatal injuries as a result of arson. These figures show that deaths relating to hospital fires are low, numbering 16 over the past 10 years, whilst non fatal casualties are significantly higher (Figure 13.3).

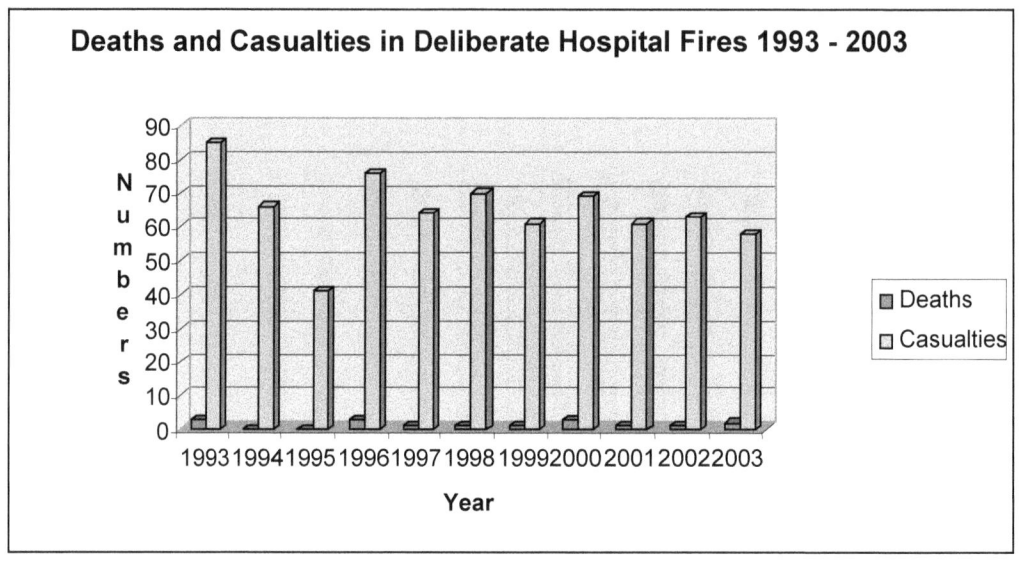

Figure 13.3 – Deaths and casualties in deliberate hospital fires 1993 – 2003
(Source: Office of the Deputy Prime Minister, 2005)

Despite the lack of data, there is sufficient evidence available to demonstrate that fire safety within healthcare premises must be a priority for NHS organisations. This is particularly relevant since a fire in a hospital poses major threats not only to patients and staff but also to the potential of the unit to continue to provide care (Roberts, 2002 p.180).

Fire safety and training at the Christie Hospital NHS Trust

The Christie hospital is a specialist cancer hospital which offers diagnosis, treatment and care for the people of Greater Manchester and Cheshire. It is the largest single site cancer treatment centre in Europe, covers a population of 3.2 million, and provides specialist surgery, chemotherapy, radiotherapy, adult leukaemia, palliative and supportive care, services for young people with cancer and endocrinology services.

Before undertaking an evaluation of fire training at the Christie hospital, I will review fire alarm reports for the hospital.

The Christie hospital maintains records of all fire alarm calls. Between 1 January 2005 and 1 July 2005, there were sixty three fire alarm reported instances. Sixty two were false alarms and one was a fire-related incident in which a patient's gown placed on a bed head light generated smoke. This was dealt with promptly by the ward staff and did not develop into a fire.

Table 13.3 shows the total number of alarm calls made between 2001 and 2005. Looking at the data, it is hardly surprising that NHS Estates (2003) has produced FPN 11 *Reducing unwanted fire signals in healthcare premises*. The impact of FPN 11 appears evident from the apparent reduction in false alarms from 2003 onwards, but there is still room for improvement.

YEAR	ALARMS REPORTED	TRUE	FALSE
2001	123	11	112
2002	110	3	107
2003	86	1	85
2004	89	0	89
2005 (Jan – July)	63	1	62

Table 13.3 – Total fire alarm calls made by the Christie Hospital 2001 - 2005

Evaluation of fire training at the Christie Hospital NHS Trust

All formal training within the Christie Hospital NHS Trust undergoes a structured evaluation process using a simple evaluation form. The results are then summarised, given a percentage score, sent to the trainer and presented annually to the Trust Board in an education report. An example of a completed evaluation summary for a fire lecture undertaken in April 2005 is shown in appendix 13.1.

A review of the training summaries between December 2004 and July 2005 showed that the average score for fire lectures was 56 percent. The induction training provided over the same period achieved an average score of 92 percent. These figures indicate that satisfaction with fire training could be improved. Figure 13.4 compares the average percentage for the evaluation of the induction and fire training sessions. This figure shows that overall evaluations of fire training range from 30 to 77 percent, whilst perceptions of induction training are much higher and range from 83 to 98 percent.

One of the evaluation questions (Question 1) asks for an assessment of the relevance of the subject in which they are being trained. These results are presented in Figure 13.5 and show that perceptions of the relevance of fire training are significantly lower than that of induction training. This is of serious concern, since this suggests that attendees do not realise the importance of fire prevention and management training. This increases the likelihood that staff attend such lectures because it is mandatory, and not because they feel the subject has relevance to them.

Comments made by the attendees at the fire lecture in April 2005 would certainly support this concern (see appendix 13.1). These comments are replicated in the evaluations from other fire lectures and these are shown in table 13.4.

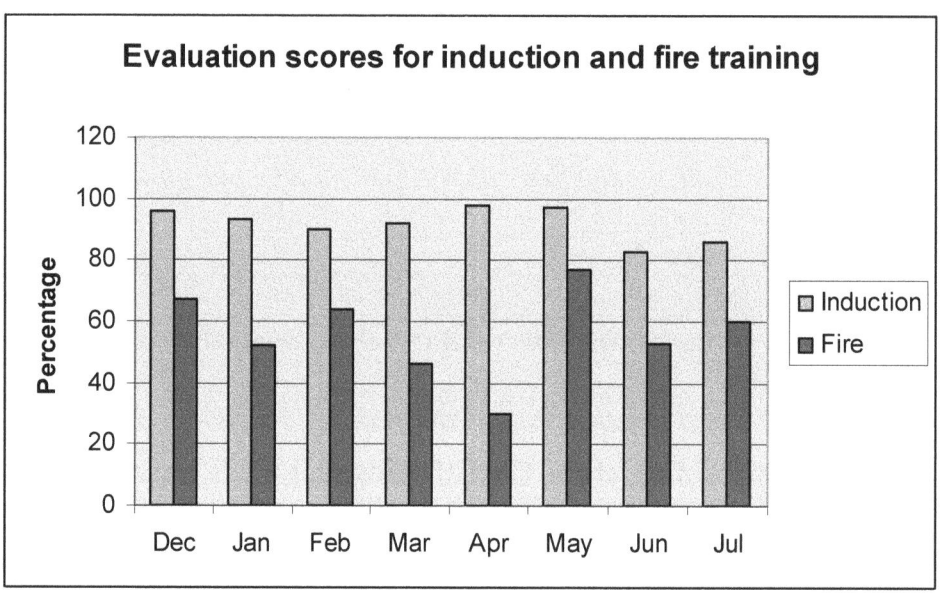

Figure 13.4 – Evaluation scores for induction and fire training

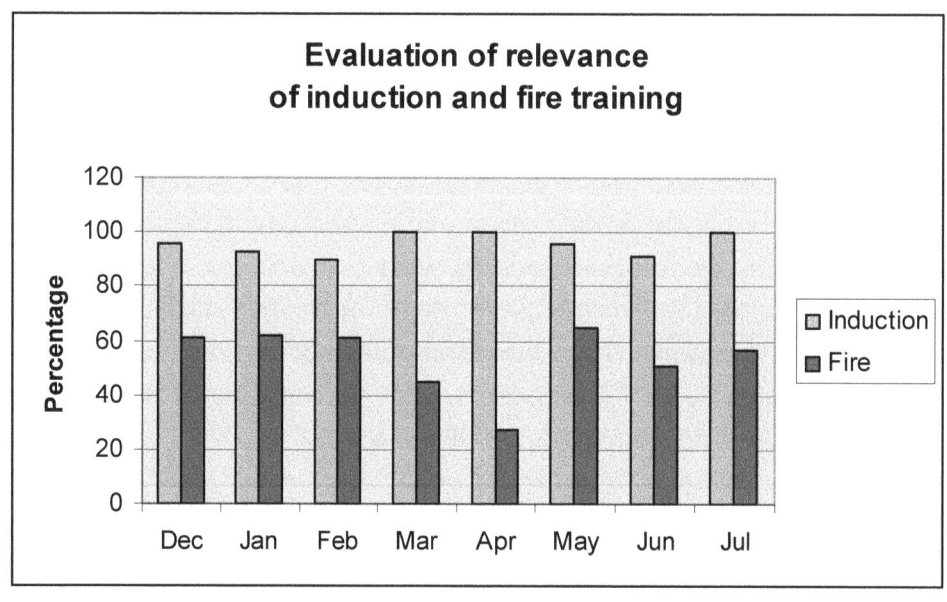

Figure 13.5 – Evaluation of relevance of induction and fire training

Positive Comments	Comments highlighting areas for improvement
• Appreciate handouts on new colours of extinguishers and what they are used for • Good visual presentation • Very interesting session. Made me realise most of fire safety is common sense. • Good to see examples of real obstructions at Christie • Good points about fire escapes and individual responsibilities in the event of a fire.	• Keep anecdote a little shorter – a lot of repetition • Still very dry, found it difficult to concentrate for whole lecture • A chart of which extinguishers can be used for different fires would be useful. • Speaker hesitated - didn't sound confident or interested. Improved as session carried on. • Need review of evacuation procedures using slings • Session on using a fire extinguisher would be useful • Handouts / intranet information may help and reinforce slides • Individual department session would be more beneficial for updated regulations • Speaking voice very quiet - could do with speaking into the microphone • On call nurse bleep holders would benefit from a session should a fire occur • A local visit to departments would help to consolidate the session • Suggestion of a quiz and some scenarios • Pace of presentation a bit quicker • Could have tried to make it more interesting - to keep attention. • Presentation could have benefited from more slides, highlighting key points.

Table 13.4 – Evaluation comments from fire lecture attendees at the Christie hospital.

To diversify and increase the depth of fire training evaluation at the Christie hospital, an additional evaluation questionnaire was devised (see appendix 13.2). This was primarily developed for the purposes of this assignment and was based on the trust evaluations which generated concerns about the effectiveness of fire training. The questionnaire was sent to 10 senior managers throughout the hospital who are responsible for a variety of clinical and non clinical areas. All 10 questionnaires were returned and all questions were answered. The results are as follows:

⊹ Department and Job Title.

The senior managers included in the survey were from the following departments:

Clinical

Haematology	Divisional Business Manager
Clinical Oncology	Divisional Business Manager
Medical Oncology	Divisional Business Manager
Surgery	Divisional Business Manager
Radiology	Divisional Business Manager

Non Clinical

Health Records	Health Records Manager
Catering	Catering Manager
Estates	Associate Director of Estates
Domestic Services	Head of Domestic Services
Pathology	Divisional Business Manager

⊹ 1. Have you attended a fire lecture in the past 12 months?

Only three Managers answered 'Yes' to this question (one clinical and two non clinical). One manager stated that she was booked for the next session in September.

⊹ 2. Why not?

Seven managers answered this question and they all attributed their non-attendance to a lack of available time. Four managers acknowledged that this was not a priority for them, since they felt they would hear nothing different to previous fire lectures they had attended.

Although not acknowledged by all seven managers, this response indicates that fire training is a low on the agenda for them all. One reason given for this is the lack of new information they feel they will hear from a fire lecture. However, there are likely to be other issues affecting this decision making process which are not evident from the responses made to this question.

3. How would you rate the effectiveness of this session in terms of:

		Excellent	Good	Average	Poor
a.	Increasing fire prevention awareness	0	0	2	1
b.	Increasing knowledge of fire management	0	0	2	1

Managers need to feel that they and their staff are using their valuable time effectively, and it is disappointing to recognise that these results indicate that this is not the situation. Feedback to peers about their perceptions of fire training will not persuade other managers to attend, or encourage them to influence their staff to see this issue as a priority.

4. What percentage of your staff have attended a fire lecture in the past 12 months?

Table 13.5 reveals the results to this question.

Department	Percentage of staff who attended a fire lecture in the past 12 months
Haematology	42
Clinical Oncology	41
Medical Oncology	39
Surgery	53
Radiology	69
Health Records	82
Catering	65
Estates	89
Domestic Services	54
Pathology	62

Table 13.5 - Percentage of staff who attended a fire lecture in the past 12 months by department

These results show that the non clinical areas have higher numbers of staff attending the annual fire lecture. Out of the clinical areas, the radiology department is most successful – and their manager had also attended a fire lecture in the last 12 months.

It is concerning to realise that so few ward areas have sent their staff for fire training. The estates department have the highest percentage of staff attending the annual fire lecture, but this is not surprising since they attend every fire alarm call in the hospital which means fire awareness is very much part of their everyday work.

> 5. Are you aware of the fire training requirements specified in the Christie fire policy?

All the managers were aware of the fire policy, but only four managers had ever referred to the policy. This question prompted five of the six managers who had never looked at the policy before to check its contents with regards training. The training section of the fire policy was regarded with scepticism since managers commented that:

1. Current training provision does **not** take account of the particular discipline of staff to be trained (as stated by the policy) - three respondents.
2. Evacuation drills have **not** taken place in the past three or four years to the knowledge of the managers (as stated by the policy) - four respondents.
3. Night staff have **never** received training aimed at testing their ability to evacuate patients (as stated by the policy) - nine respondents.
4. Individual training requirements have been highlighted in the past, but have **not** been addressed (as stated by the policy) - two respondents.
5. All switchboard staff have **not** received specific training despite receiving particular mention in the fire policy - one respondent.

These responses reflect the discontent felt by managers over the current provision of fire training, particularly when the policy describes what should be happening and it is obvious, in their opinions, that this is not the case.

> 6. Have you assessed the risk of fire in your department in the last year?
> Yes | 10 | Go to question 7a
>
> No | 0 | Go to question 7b

> 7a. YES – what is the score and how has this been calculated?

All managers had calculated the risk assessment score for fire in their departments, using a 5 x 5 risk assessment matrix (see table 13.6).

	Consequence				
	1	*2*	*3*	*4*	*5*
Likelihood	*Insignificant*	*Minor*	*Moderate*	*Major*	*Catastrophic*
1 – Rare	1	2	3	4	5
2 – Unlikely	2	4	6	8	10
3 – Possible	3	6	9	12	15
4 – Likely	4	8	12	16	20
5 - Almost Certain	5	10	15	20	25

Table 13.6 – Risk assessment 5 x 5 matrix

Using the matrix is straightforward. Likelihood refers to the chance of the risk happening whilst consequence relates to the actual impact of the risk. The scores are allocated a risk classification by colour at the Christie hospital as follows:

- **Green** category risks score between 1 and 3 and are deemed acceptable to the Trust (Low risk).
- **Yellow** category risks are classified as moderate risk and score between 4 and 7.
- **Orange** category risks score between 8 and 14 and are classified as significant risk.
- **Red** category risks are classified as high risk, scoring between 15 and 25.

The risk assessment scores and risk classification for fire in the respondent's departments, using a 5 x 5 risk assessment matrix, are shown in table 13.7.

Department	Fire risk assessment score	Classification
Haematology	4 x 4 = 16	**HIGH**
Clinical Oncology	3 x 4 = 12	**Significant**
Medical Oncology	4 x 4 = 16	**HIGH**
Surgery	4 x 4 = 16	**HIGH**
Radiology	3 x 5 = 15	**HIGH**
Health Records	3 x 5 = 15	**HIGH**
Catering	4 x 5 = 20	**HIGH**
Estates	4 x 5 = 20	**HIGH**
Domestic Services	3 x 4 = 12	**Significant**
Pathology	3 x 4 = 12	**Significant**

Table 13.7 – Fire risk assessment score and classification by department
(Based on likelihood x consequence = score)

The risk assessment scores presented in table 13.7 vary between 12 and 20, which leads to a risk classification of high for seven departments and significant for the remaining three areas. This would indicate that the risk of fire is of concern and a priority area of action for the managers to address.

✦ 7b.   If NO – why not?

This question was not answered, since all managers had assessed the risk of fire in their department over the last year.

✦ 8a Has the risk of fire been added to your departmental risk register?

   Yes | 10 | Go to question 8b

   No | 0 | Go to question 9

✦ 8b. Please describe the action plan that has been developed on your risk register to deal with the risk of fire.

The transfer of risk assessments to the risk register is an essential aspect of the risk management process and it was gratifying to find that all managers had transferred their risk assessment for fire onto their departmental risk registers. However, their action plans were all similar and were based on risk register training they had received in the past year, where fire was one of the examples they had used in practice.

All the action plans included the need for staff to attend annual mandatory training and the completion of the monthly maintenance audit programme which addresses fire issues such as checking the availability of fire extinguishers and ensuring fire exits are kept clear. However, no manager had personalised their risk register for their own department, especially for a risk that had been assessed as being so important to address. This suggests that managers have not fully integrated the risk and management of fire into the culture of their department.

- 9. Do you feel the current fire training sessions reflect the risk of fire in your area?

This question was answered negatively by all the managers. three respondents added further comments:

"I have been disillusioned with fire training in the NHS for many years and acknowledge that I do not encourage my staff to attend annual fire training – I feel it is a waste of time, boring and repetitive. However, I am also aware that the risk of fire must be taken seriously and this whole issue is a real dilemma for me"

"I attend a fire lecture every year and encourage my staff to do the same, but genuinely feel it is time wasted and only go because it looks good for my department. The training provided does not relate to my department and I doubt it would help any of us if there really were a fire."

"I have calculated fire as a high risk for my department – the fire training provided would be suitable if this were a low risk as the training is low key and unhelpful"

The responses to this question reflect the frustrations felt by departmental mangers who have all identified fire as a high or significant risk, but do not feel the current training provided meets this level of risk.

- 10. How do you feel the content and delivery of fire training could be improved to make it more effective and relevant to your department?

Every manager responded to this question with suggestions, which can be interpreted as an indication of their concerns over this issue. In addition, the number and variety of suggestions, listed in table 13.8, demonstrate the thought given to this question.

1.	Don't show videos of ancient hospital fires – use videos of fires that are more recent and are meaningful to get the fire safety message across. For example, the Bradford fire highlights the speed and potential severity of fires.
2.	Make the training session more interactive – to encourage discussion and increase learning.
3.	Make use of online learning packages. There are several around and they can be individualised to specific wards and hospitals.
4.	Invite the fire officer to ward and departmental meetings and have a question and answer session to ensure staff know, for example, where the fire alarms and extinguishers are located, where the fire exits are. The key thing is to raise the profile of fire awareness.
5.	Hold different sessions for different staff groups and adapt the session to the audience, instead of repeating the same old thing irrespective of who is listening.
6.	Some staff will have more responsibility in the event of a fire, for example, the bleep holder in charge of the hospital; and these staff should receive individualised training to ensure they are clear about their responsibilities.
7.	Ensure all new staff, including bank and agency staff are aware of the different fire arrangements in each area as soon as they start work. This would probably be best done by providing a map of the department with alarms, fire exits, etc highlighted and a brief summary of what should be done in the event of a fire.
8.	Plan to undertake practice evacuation drills as specified in the fire policy – this will bring the situation to life and make the process more meaningful.
9.	Carry out fire training in wards and departments, discussing fire prevention and management on site; and consider what would happen if a fire broke out in a specific area – this would be better than always having the training session in the auditorium.
10.	Make the most of the opportunity to talk about fires at home as well – we all have lives outside of work and more people are killed in house fires than hospital fires. We need to change society's attitude towards fire – and we are part of society as well as a workforce.
11.	Take key staff off site and visit the local fire brigade to hear from them about fires they have dealt with – this will help to raise the profile of fire prevention.
12.	How about working with a fireman for a day? This would increase knowledge about fires and also increase my morale!!

Table 13.8 – Suggestions from managers to increase
the effectiveness and relevance of fire training

Some of the suggestions presented in table 13.8 reflect the content of the fire policy, indicating that if what was planned actually did take place, then this would be beneficial. Other comments are more radical and some more achievable than others, but the ideas proposed generate interest and enthusiasm for the future of fire training.

Interestingly, the respondents did not challenge the provision of mandatory annual fire training for themselves and their staff. Consequently, they did not take the opportunity to debate whether or not it might be more effective to consider that some staff should receive fire training less often, whilst others should participate more often.

These ideas (and others) will now be discussed in more details as a proposal for the restructuring of fire training in the health service is developed.

Proposed restructuring of fire training in the health service

The following proposal is based on the evidence collected for this chapter and focuses on five key questions:

1. Who?
2. What?
3. Where?
4. When?
5. How?

1. Who?

The title and responsibilities of the person delivering fire training within NHS organisations varies considerably. In addition, there is also notable variety in the background, experience and qualifications of these fire trainers – although it is common for the trainer to have had some previous experience in the fire service.

It is proposed that a national personal specification be developed which would establish the essential criteria for the role of the fire trainer. This would include experience and knowledge of fire fighting and an educational qualification. This would help to ensure that appropriately qualified people with an explicit interest in teaching as well as fire management are appointed to these roles. It is also proposed that the job title and responsibilities for fire training are standardised so that differences between organisations are reduced and the highest quality of training is provided.

It is further proposed that the qualified fire trainer is supported by a nominated member of the fire brigade who is an active fire fighter. This would help develop links between the fire service and the healthcare organisation, and could increase interest in, and attendance at, the mandatory fire lecture.

2. <u>What?</u>

The fire lecture should address the essential areas of the causes of fire, fire prevention and fire management. In particular, the training should ensure staff know how to call for help if a fire is suspected, where the fire alarms are and how these should be activated, where the fire escapes are and how to keep them available for use at any time, where the fire extinguishers are and when they should be used and how to evacuate the premises. These subjects are currently covered in some format in the present fire training format and are important areas to retain. However, it is proposed that the opportunity of fire training at work is also used to discuss the prevention and management of fires in the home, on holiday and whilst socialising, since this is when most people are at risk from fire.

3. <u>Where?</u>

All fire training currently takes place in a formal setting such as a lecture theatre, although some fire trainers do venture out into departments to discuss the risk of fires in these areas.

It is proposed that there should continue to be a formal fire training session, since this will allow a large number of people to be trained in one session. However ward or departmental fire training should be compulsory, since this will allow healthcare workers to think about the risk of fire in their own areas and develop appropriate and effective action plans. It is also proposed that off site training should be provided to encourage people to increase their interest in fire prevention and management, and that this should take place at the local fire station. It is important that a culture of fire safety is developed, and raising its profile in this way will be beneficial.

4. <u>When?</u>

Most organisations provide fire training on induction for all new starters. However, it is proposed that this arrangement should be mandatory, so that all new starters **must** receive fire training, including temporary staff. Providing an annual update 10 months after a person has commenced work would not be helpful if a fire had occurred during their first month.

Although legislation requires fire training to be provided annually, it is known that many staff do not receive this frequency of training. In addition, there are doubts that such training is required for all staff. For example, a clerical worker whose responsibility in a fire would be to safely leave the premises does not need the same training as a ward manager who would be required to take responsibility for a number of patients and other members of staff.

It is therefore proposed that staff members who have minimal responsibility in the event of a fire should receive formal fire training on induction, a departmental fire training session within the first year of employment and three-yearly thereafter, unless the department alters its location or the staff member's responsibility changes. Those staff who have significant responsibility in the event of a fire should continue to receive annual training, but this should cover the specific responsibilities of their role. To ensure this proposal is effective a training needs analysis would be required for each post holder to calculate their specific training requirements in respect of preventing fire and/or responding to a fire.

5. How?

A number of suggestions have been made in this chapter to increase the effectiveness of fire training. It is proposed that the current fire training arrangements should be restructured to include an increase in the number of departmental or ward based fire training sessions. These sessions should include:

- Practical sessions, for example: locating fire alarms, fire extinguishers and fire exits.
- Walk through evacuation drills, discussing problems as they are highlighted.
- A locally developed quiz to check staff are fire safety aware.
- Encouraging staff to learn through question and answer sessions with the fire trainer and other key staff.
- Discussion of specific scenarios which are relevant to that location.
- Maps of the department which identify the location of fire equipment and emergency exits.
- Discussion of actual fires, what went well and what could have been done differently so that the significance of fire is highlighted.
- Opportunities for on line learning or the use of videos and compact discs which have been developed for the hospital and specific departments.

To facilitate the usefulness of this proposal, it is important that the wards and departments are offered as many sessions as necessary to cover all staff over a given period of time.

Conclusion

This chapter has shown that training of healthcare staff in fire prevention and management is not based on research and is often not educationally sound. Fire training is currently provided by fire advisors who may not have teaching qualifications. However, they do have, on the whole, a wealth of experience and knowledge of dealing with fires. This knowledge and experience must not be ignored, but it must be possible to make the training process more robust by ensuring that fire advisors undergo a formal teaching course of some description.

The basis for the content of the fire lectures is legislation and there has been very little (if any) research undertaken to assess the effectiveness of fire training in the past. Encouraging fire advisors to undertake teaching courses may generate the type of questioning required to challenge legislation and develop more effective methods of delivering fire prevention and management training to healthcare staff.

This chapter has also highlighted a number of ways in which fire training might be made more meaningful and productive in the future – and, as shown through the evaluation of fire training at the Christie hospital, there can be little doubt that a thorough overhaul of the current process is required.

Appendix 13.1 - Completed fire training evaluation proforma

POST EVALUATION OF TRAINING

Date of Training	6th April 05
Trainer	
Training Undertaken	Mandatory
No of Delegates	29

		Scoring
1	Relevance of subject to the delegates attending	27.78%
2	Presentation delivery & effectiveness of Tutor	30.56%
3	Quality of training materials	32.64%
4	Standard of Training Facilities	31.94%
	AVERAGE	30.73%

Delegate Comments

1	Disjointed presentation at first but improved as the session went on
2	Slide show very informative and easy to follow
3	Could have done with a video tackling fire
4	More practical information, what to do in fire & evacuation procedures. Kept short easier to remember
5	Too much detail on legislation. Delivery too slow, not specific to hospital fire safety.
6	Would like training on fire slings
7	Speaker needs to be more concise, to retain attention of the audience
8	Would like some ward based training
9	
10	

Trainer Comments

ACTION

Appendix 13.2 - Fire training evaluation questionnaire

FIRE TRAINING EVALUATION QUESTIONNAIRE

Department:
_____

Job Title:
_____

1. Have you attended a fire lecture in the last 12 months?

    No ☐ Go to question 2

    Yes ☐ Go to question 3

2. Why not?
    _____

3. How would you rate the effectiveness of this session in terms of:

		Excellent	Good	Average	Poor
a.	Increasing fire prevention awareness				
b.	Increasing knowledge of fire management				

4. What percentage of your staff have attended a fire lecture in the past 12 months?
    _____

5. Are you aware of the fire training requirements specified in the Christie Fire Policy?
    _____

6. Have you assessed the risk of fire in your department in the last year?

    Yes ☐ Go to question 7a

    No ☐ Go to question 7b

FIRE TRAINING EVALUATION QUESTIONNAIRE, continued.

7a. If YES – what is the score and how has this been calculated?

_____

7b. If NO – why not?

_____

_____

8a. Has the risk of fire been added to your departmental risk register?

    Yes   ☐   Go to question 8b

    No   ☐   Go to question 9

8b. Please describe the action plan that has been developed on your risk register to deal with the risk of fire.

_____

_____

9. Do you feel the current fire training sessions reflect the risk of fire in your area?

_____

_____

10. How do you feel the content and delivery of fire training could be improved to make it more effective and relevant to your department?

_____

_____

Thank you for taking the time to complete this questionnaire – please return to _____

References

Ashford and St Peter's Hospitals NHS Trust (2003). *Fire policy and procedural notes.* www.ashfordstpeters.nhs.uk

Christie Hospital NHS Trust (2004). *Risk management strategy.* www.christie.nhs.uk/Publications/docs/pdf/risk_management_Strategy.pdf

NHS Estates (1994a). *Firecode policy and principles.* www.doh.gov.uk

NHS Estates (1994b). *HTM 83 - Fire safety in healthcare premises - general fire precautions.* www.doh.gov.uk

NHS Estates (1994c). *HTM 86 - Fire risk assessment in hospitals.* www.doh.gov.uk

NHS Estates (2003). *Reducing unwanted fire signals in healthcare premises.* Fire Practice Notice 11. www.doh.gov.uk

NHS Estates (2004). *Firecode.* www.doh.gov.uk

Norfolk and Norwich University Hospital NHS Trust (2004). *Fire safety policy.* www.nnuh.nhs.uk/docs%5Ctrustdocs%5C16.pdf

Office of the Deputy Prime Minister (2005). *Fire statistics, United Kingdom, 2003.* [Now the Department for Communities and Local Government] www.communities.gov.uk/index.asp?id=1124891

Roberts G (2002). *Risk Management in Healthcare.* (2nd edition). Witherby & Co. Limited, London. www.witherbys.com

Roberts P (2003). *Managing fire safety in the NHS in England.* Business Briefing: Hospital Engineering and Facilities Management. 1-3. www.bbriefings.com/pdf/13/hosp031_r_proberts.PDF

Royal Devon and Exeter NHS Foundation Trust (2004a). *Fire precautions policy.* www.rdehospital.nhs.uk/foi/classes/docs/class14/Fire%20Advisors%20Policy.pdf

Royal Devon and Exeter NHS Foundation Trust (2004b). *Trust fire advisors annual report.* www.rdehospital.nhs.uk/foi/classes/docs/class14/Fire%20AdvisorsAnnual%20Report.pdf

United Bristol Healthcare NHS Trust (2004). *Meeting of the Trust Board – controls assurance assessment.*
www.ubht.nhs.uk/board/meetings/2004-05/May/Controls%20assurance%20board%20report%20-%20May%202004.doc

United Lincoln Hospitals NHS Trust (2005). *Risk assessment policy (draft).*
www.ulh.nhs.uk/TrustBoard/2005/March/Supporting_Papers/Enc%20D%20Risk%20Assessment%20Policy%20Draft%2014-3-05.pdf

This page is intentionally blank.

14

The risks and opportunities presented to the NHS by the disposal of surplus buildings

SHIRLEY MUNDAY

Introduction

The NHS property portfolio is the largest estate in Europe. It absorbs over 20% of NHS running costs (NHS Estates, 2000) and is replete with risks, and opportunities.

When it was established in 1948, the NHS inherited a rich legacy of healthcare buildings. Some were historic buildings in historic landscapes and many make a significant contribution to the character of historic towns. As technological advances are made and methods of clinical care are changing, so many of these buildings are no longer fit for purpose.

NHS Trusts were established in England from 1991 onwards. Only properties thought to be of long term use were transferred over to each Trust. The remaining premises were known as 'retained estate'. When 'care in the community' initiatives were established after 1990, many hospital services were merged and several institutions closed. This led to the sale of approximately 150 former psychiatric hospitals and hospitals for the mentally handicapped, many of which were built originally as Pauper Lunatic Asylums or work houses.

Also in the early nineties, when Crown Immunity was fully removed from the NHS, NHS Trusts became 'landlords' and any residential properties they had had to comply with various statutory requirements relating to housing, fire, etc. As most buildings were old, and many were in a poor state of repair and failed to meet statutory requirements, many hospitals had no choice but to review how their accommodation was utilised and seek alternatives. This resulted in many properties being sold to housing associations, to the private sector, or being demolished.

The 'surplus' NHS estate

The size of the 'surplus' NHS estate is substantial and disposals potentially provide valuable additional resources to fund NHS developments. It is therefore important that sales are conducted in a way that achieves the best value.

The Select Committee on Public Accounts (1999) examined the management, utilisation and condition of the NHS estate across the NHS in Scotland. They identified concerns over the time taken to identify and dispose of surplus land and buildings. More importantly, they were "very concerned that eight years after Crown Immunity was removed from the NHS estate, 29 per cent of it did not comply with safety and statutory standards, and that the trusts involved were breaking the law."

The Health Act (1999) and The Health and Social Care Act (2001) Section 45 provided for the establishment of Primary Care Trusts (PCTs), identifying that, as is the case for NHS Trusts, they would have the power to acquire, own, develop and sell property and be accountable for the management of their estate from the day they become operational. Estate portfolios were to be kept under regular review and any surplus estate was supposed be disposed of as soon as possible. If the value of the property was over £1 million, then approval had to be obtained from NHS Estates, who acted on behalf of the Secretary of State for Health.

Her Majesty's Treasury, sets out procedures for dealing with surplus property owned by the Government which includes Ministry of Defence, government buildings and the NHS estate.

A joint working party was set up with English Heritage and NHS Estates (an executive agency of the Department of Health) and a report was commissioned by the NHS Executive at the Department of Health (NHS Estates, 1995). The report, titled *Historic Buildings and the Health Service*, established guidelines for all levels of management who may seek advice on listed buildings and conservation matters, including the problems surrounding the care and maintenance of historical buildings, and disposal, and potential for conversion to new uses.

NHS Estate's *Estatecode* provides guidance to the NHS on managing their estate, including statutory requirements relating to the ownership and management of land and property and the DH mandatory requirements including property transactions and commercial requirements. This guidance complements and should be followed in conjunction with NHS Estates (1999) *Developing an estates strategy.*

In 1998, as part of its comprehensive plan to improve efficiency and productivity in the public sector, the government recruited a team of top private sector managers to the Public Services Productivity Panel (PSPP). The Panel worked collectively and individually with government departments and agencies to drive out waste and make public money work harder. A project team consisting of representation from NHS Trusts, the NHS Executive (part of the Department of Health) and NHS Estates produced a document called *Sold on Health* (2000). The aim was to deliver improvements in the efficiency and effectiveness in the procurement, operation and disposal of NHS estate and to do so whilst having due regard to the wider interests of the Government. The report recognised that the value and volume of disposals was increasing and that few trusts had the experience or skills to manage the surplus estate in maximising its development and income potential.

NHS Estates (1999) *Developing and Estates Strategy,* issued guidance on rationalising NHS estate. However, the NAO (2003) identified gaps in the guidance on disposal, for example,

- identifying the cost of holding surplus property including maintenance, and security
- Assessment of suitable disposal for example demolition versus refurbishment

The National Audit Office (2002) has carried out extensive research on the management of property and its disposal in the NHS. This report comments on the good practice and highlights gaps where improvements can be made. It supported guidance given by the Estatecode and supported the working partnership with English Heritage and encouraged close liaison with local planning authorities to speed up the sale process.

Government (2003) *The Communities Plan* (Sustainable Communities: building for the future) makes reference to the setting up of a register of surplus public sector land. The key objective of the Register is to provide a central database of surplus public sector land. The intention is for public bodies to be able to easily identify surplus land that can be transferred within the public sector, in order to improve public service delivery, prior to it being placed on the open market for sale. NHS Bodies are encouraged to provide details of their surplus estate for inclusion on the Register. Priority purchasers remain first priority if another public sector body expresses an interest alongside the NHS.

The Valuation Office, an executive agency of Her Majesty's Revenue and Customs (HMRC) provided information on the best value from disposal of surplus property in the Secretary of State's ownership. Transfer of accountability for disposal of surplus NHS property to NHS Estates in 1999 allowed a national overview.

The National Audit Office (2002) found that in 2000 NHS Trusts owned some 95 % by value of all land and property (8,750 hectares by area) in the NHS in England. This was valued at £23 billion, with a replacement value of £76 billion. By comparison, the value of the NHS estate in Scotland was estimated at some £3 billion (Select Committee on Public Accounts, 1999).

1994/95 saw the fourth major wave in the creation of NHS Trusts and the retained estate at that time had an estimated value of £1.2 billion, NAO (2003) *The Management of Surplus Property in the NHS*. NHS Estates since then has conducted a programme of disposal exceeding targets agreed with the NHS Executive.

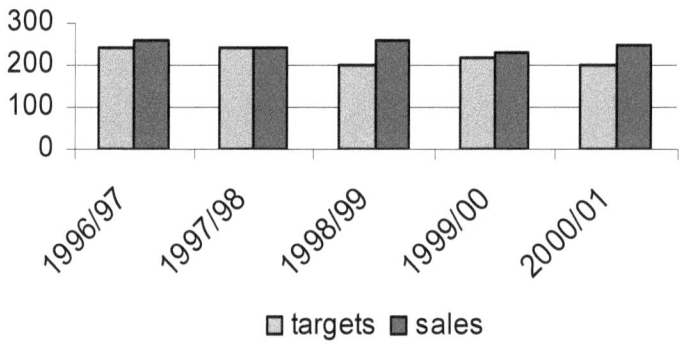

Figure 14.1 - Source NHS Estates cited in National Audit Office (2003)

The National Audit Office (2003) identified that trusts obtained at least £380 million from the sale of surplus property from 1997 to 2000, with plans to sell property worth over £700 million between 2000 and 2003. This report also cited that the valuation office identified that there were 1000 non-operational sites with an estimated market value of £912 million, of which 258 were held by NHS trusts.

**Identified Surplus Properties for sale
2000 - 2003**

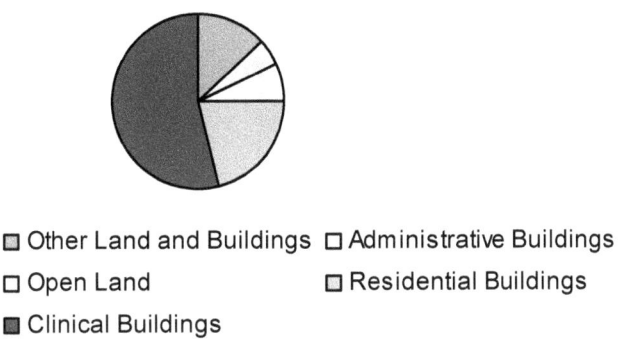

☐ Other Land and Buildings ☐ Administrative Buildings
☐ Open Land ☐ Residential Buildings
■ Clinical Buildings

Figure 14.2 - Source National Audit Office (2003)

HM Treasury (2002) provides guidance on disposal of surplus land and property in the public sector. They also hold a register of Public Sector Land. Once land is placed on the register other public bodies, including the NHS, have 40 days in which to express an interest in purchasing. Any subsequent transfer is at market value. If no interest is shown in that time the property can be sold on the open market. This guidance (see annexe 24.2) also advises on cases for the delay in sales, for example

- ❖ if the market is flooded with properties at that time or if a higher price can be obtained by selling several properties at the same time.
- ❖ There may be occasions where a short term lease would be appropriate if there is little prospect of an early sale.
- ❖ In cases where former owners or sitting tenants are a consideration the "Crichel Down rules" apply (Office of the Deputy Prime Minister, 2000). Where tenants have been sitting tenants for a number of years (more than two) and have carried out improvements to the property, the department may wish to consider sympathetically any offer from such a tenant..

NHS Estates in their *Sold on Health* (NHS Estates, 2000) report recognised that the value and volume of disposals was increasing and that few trusts had the experience or skills to manage the surplus estate maximising its development and income potential. Figure 14.3 shows information on actual and anticipated sales of surplus estate 1995/96 to 2001/02.

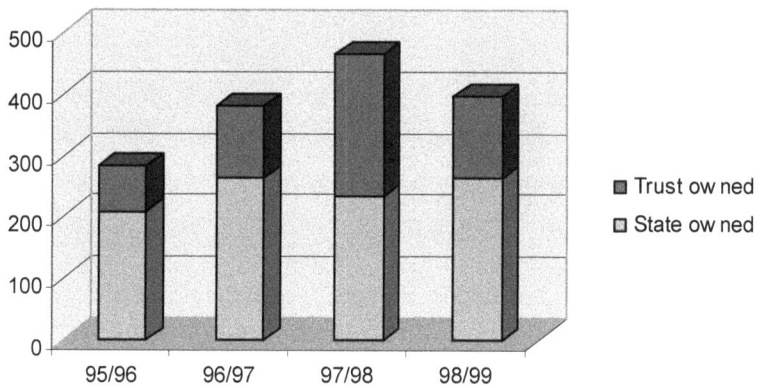

Figure 14.3 - Source: NHS Estates (2000)

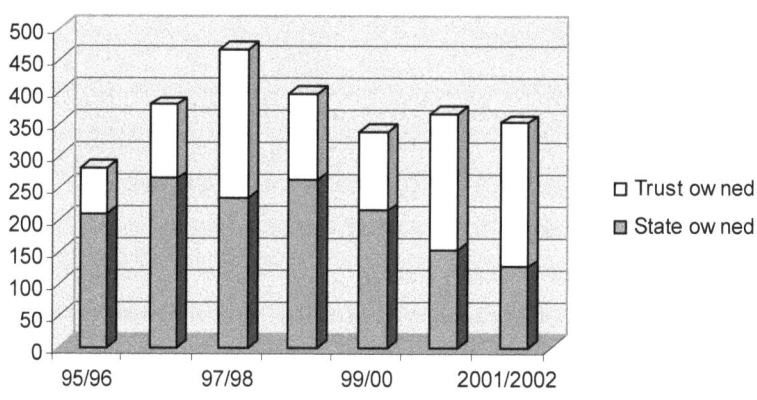

Figure 14.4 - Source NHS Estates (2000)

The first recommendation from *Sold on Health* (NHS Estates, 2000) was that "there should be a national framework and regional overviews for estate within the NHS, to cover procurement, operation and disposal. These would need to be supported by local estate strategies, as part of health improvement programmes, which would be reviewed and amended to reflect local need and modernisation of the NHS."

In the NHS, estate not allocated to Trusts remains in the Secretary of State's ownership, and is managed and disposed of by NHS Estates. *Sold on Health* identified that ownership of 95% of the NHS estate was with more that 400 Trusts. This number has grown as primary care trusts were established. Recommendation number 3 with regard to disposal of surplus property states that "Disposal of all estate will need to be overseen corporately by NHS Estates, acting as the informed client, for trusts as it does for the NHS Executive to maximise proceeds. The disposal process will be managed by public or private organisations which would be subject to best value or market testing to ensure value for money is achieved."

The National Audit Office (2003) reported that most of the properties in the £400 million retained estate portfolio after 2001/2002 were subject to sale through a Public Private Partnership Initiative. This followed recommendation number 4 in *Sold on Health* that "The opportunity to have a one off disposal of the surplus estate through a public/private partnership should be explored in greater detail, while continuing the ongoing disposal process."

One of the principal reasons for delay in disposal of historic buildings is the local planning process, as identified in Figures 14.5 and 14.6. In the case of Figure 14.6, it can be seen that major investment in property redevelopment is mainly tied up in bureaucracy. In May 2004 a conference was held to discuss *Re-use of historic hospitals: health v heritage*. The Senior Property Advisor at NHS Estates pointed out, at the conference, the difficulties in finding buyers with viable alternative uses to satisfy local planning and statutory requirements for listed buildings. A plea was made to local planning authorities to help speed up the process by addressing political agendas resistant to modernisation and change.

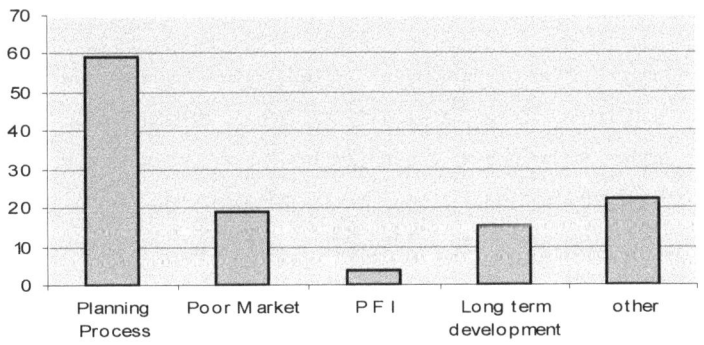

Figure 14.5 - Source: NHS Estates (2000)

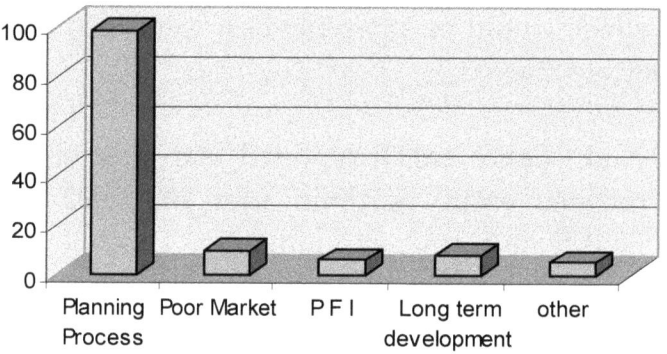

Figure 14.6 - Source: NHS Estates (2000)

The process of disposal

Estatecode (NHS Estates, 2003) identifies the process that trusts must follow before disposing of their assets. Account must be taken of the healthcare delivery plans and there should be periodic reviews of the estate to identify surplus sites. Also, consideration must be given to the rights to buy that may exist to former owners of the property (Department of the Environment, 1992).

The NHS Executive set out guidance on Private Finance Initiatives in relation to selling land (NHS Executive, 1999). NHS Estates (2000), in their *Sold on Health* publication, recommended that surplus land not integral to the development should be excluded from PFI procurements, in agreement with the circular.

NHS Estates (2000) *Sold on Health* recommends that a Trust or Health Authority would have to inform their Regional Head of Estates and Facilities as soon as a property becomes surplus to requirements. At this point, a decision would be made as to whether the Trust has the local expertise to manage the proceeds or, if the sale is too complex, the Health Authority (referred to as the 'informed client') would recommend that the sale be managed externally.

The role of the 'informed client' is to:

- Provide advice and guidance
- Assist in the selection of external advisors
- Agree strategic considerations
- Review progress ensuring relevant guidance has been followed

Many properties in the NHS are specialist in nature and there is usually no market to sell them on in their existing use. To ensure maximum sale value is achieved, alternative uses must be investigated. This may involve planning and listed building applications, possible green belt applications and dialogue with local residents and pressure groups as well as local and national politicians. This will inevitably delay the sale but should maximise the sale price.

The following is a list of key stakeholders that might contribute to the sale process:

- Department of Health
- NHS Estates
- NHS Trusts
- Advisors
- Buyers
- Department for Transport
- English Heritage
- Local Planning Authorities

Potential Risks

A number of potential significant risks exist in the process of disposing of NHS property, including:

- **Fraud** - HM Treasury (2000,) 28.3 highlights the possibility of fraud on transactions, and gives advice on the separation of duties, i.e. to ensure that no one person is able to control all aspects of payment and receipt. It also gives advice on frequently changing encrypted passwords to electronic systems

- **Disposal of Gifts or Charitable Assets** - HM Treasury (2000) *Government Accounting* "where the assets of a charitable trust are disposed of, it must be for the trustees to decide what use is made of those assets within its charitable objectives. It would be improper for the trustees to hand over the money to someone else to decide how to use it." This report goes on to cite a case *Liverpool and District Hospitals v. AG (1981) Ch 193*, "The court held that the company was the legal and beneficial owner of the assets, but was in a position analogous to that trustee and that the assets of the charity could only be used to further the charitable objectives of the company. As such, only it could decide how to use the assets". The problems arise when the charity has built up their assets through grants and in grant aid.

- **Sitting Tenants** - Under the *Crichel Down* rules (Office of the Deputy Prime Minister, 2000 and HM Treasury), a person will be regarded as a sitting tenant if there is a regulated tenancy under the Rent Act (1977) and that tenancy commenced before January 1989, unless it was an Assured Shorthold Tenancy (fixed Term for at least six months). In the context of the rules a "sitting tenant" was intended to apply to tenants with indefinite or long term security of tenure. This could cause considerable delays in a potential sale.

A case study involving delay in selling NHS property

The BBC (5 July 2002) reported that the NHS was losing money over property sales. Their report highlighted an NHS trust in Wales that had been criticised for losing money from the sale of properties. A case was cited involving the sale of a former psychiatric hospital in Denbigh. Built in 1840s, this site was identified surplus to requirements in the mid 1980s. After a period of consultation in 1991, the site closed down in 1995. A number of offers of around £1m were received between 1996 and 1998 and, following several years of delays, the listed site was sold for £155,000 in 1999.

The BBC also quoted Sir John Bourn (the Comptroller and Auditor General for England and Wales at the National Audit Office) as saying "If the NHS Trusts and health authorities had halved the overall time taken in 1999–2000 and 2000–2001 they could have saved the NHS in Wales some £1m in related disposal costs."

Other Risks

There are many risks and hidden costs associated with empty properties, which can be very costly to the NHS. These include:

- Under The Occupiers Liability Act (1957), all Trusts have a duty of care to ensure that the property is maintained and poses no danger to members of the public, including any unauthorised person who may enter the premises. The hidden costs incurred when discharging this duty are for security, building maintenance and inspection.

- Capital charges occur when providers of public services must pay for their capital through the mechanism of an annual charge based upon the value of assets used in service provision (Heald and Scott, 1996). Depreciated Replacement Costs (DRC) are difficult to assess when buildings are listed and there is a possibility that capital charges on surplus properties will be incorrect. However if a building has been zero rated (and written off the books), capital charges would not applicable.

- Property prices usually rise as years go by and there may be a temptation to wait until the market is more favourable. However, empty premises are at risk from vandals who could damage or set fire to a property, which in turn will devalue the premises, and make them less attractive on the open market.

- Potential for the property to deteriorate therefore will lose value over time and possibly reduce the attraction to future buyers.

Case study on the disposal of an NHS building

At Rotherham NHS Foundation trust, one building had been out of use for three years and concerns had been raised both about safety and security as well as the facts that valuable land inside the hospital complex was not being utilised. The building had been zero rated for value and £1.1million pounds had been written off the assets of the trust. It was decided that the building should be demolished. The actual building, known as Beechcroft, had no features of special interest. It was constructed of brick and breeze block cavity wall with a concrete floor and a flat roof, therefore there was nothing that could be considered for salvage or reclamation.

Potential uses after demolition had been identified as a possible location for the building of a new unit that could be leased out to the Mental Health Department and generate income for the Trust, or for the creation of additional office space, or for use as a car park. It was decided to create a car park

The Process

Step 1 - It had been decided that the building was beyond economical repair and surplus to requirements therefore NHS Estates were informed.

Step 2 - A tendering exercise was carried out that included demolition of the building and building a new car park. The best price came in at £1million. This was too expensive so an alternative option was sought.

Step 3 - A second tendering exercise was carried out that included demolition but no car park. The cost of this came in at £314,000. Due to financial constraints within the Trust, this was the preferred option. This cost included termination and diversion of services to the building. As part of the demolition process, the contractor offered a 'cut and fill' service. This meant that a wet crusher would be brought on site (to minimise the creation of dust) and all the rubble, concrete and bricks would be crushed and used as hardcore to level the land after demolition.

Step 4 - In the cellar is a plant room that feeds hot water to other buildings. When a survey was carried out, asbestos fibres were found from a previous removal. The cost of removing asbestos from the whole cellar was £100,000; however the cost partial removal that is, from where the plant was and then sealing off the remainder was £45,000 to £50,000. This also delayed the start of the project by four weeks.

Step 5 - Terminate services. The fire alarms and the hot water were connected to another building. The gas connection had to be terminated and oil tanks removed. In the cellar, there was also a building management system that controlled services to other buildings. The telephone links, IT links and electricity also had to be disconnected.

Step 6 - Commence demolition.

Discussion

The NHS estate had an estimated value in 1998/99 (National Audit Office, 2003) of £23 billion with a replacement value of £76 billion. The programme of disposal from 1997 to 2000 generated an income from sales of at least £380 million, with an expected income of £700million from 2000 to 2003, in addition to 1000 non-operational sites with a market value of £912 million. There appears to be little evidence to show exactly what has happened to the money that has been generated (a potential of £2 billion between 1997 and 2003).

Until NHS Trusts are established as Foundation Trusts, they will have a problem reinvesting the income from sales wisely, as surplus money in accounts at the end of a financial year cannot be carried over to the next financial year. Therefore any substantial income would have to be spent before the end of the financial year, leading to opportunities for capital investment in large projects being missed.

An example of the consequences of lack of reinvestment can be demonstrated with staff accommodation or hospital residences. When many hospitals were built, it was recognised that the provision of accommodation for staff was essential as there was a need to provide affordable housing for students, newly qualified staff and doctors on rotation. Staff were charged a nominal rent and in many cases received additional abatements according to their on-call commitments. In some cases accommodation was provided free of charge, e.g. for medical students, pre-registration house officers and doctors working a one in three on-call.

When Crown immunity was fully removed in the early nineties, many of the residences in the NHS had fallen into disrepair, or failed to meet relevant statutory requirements. Generally speaking, income from rents was never reinvested in the accommodation, but went into the 'general pot' for Estates and Facilities. Staff accommodation had, historically, been low priority as far as re-investment was concerned and only essential maintenance was carried out. Therefore, when a Trust became a Landlord, the cost to refurbish or maintain accommodation played a large part in deciding what to do next. Many Trusts decided to sell off their accommodation to housing associations, which did at least leave a supply of housing available for staff, but at rents that were much higher than hospital rents. Others turned their accommodation into car parks or offices.

This has left the legacy that some hospitals, especially in the South of England and inner city areas, can no longer recruit to lower grade posts, as the cost of housing to buy locally is unaffordable to key workers. In addition to these pressures, the British Medical Association (2001) insisted on minimum standards of accommodation for junior doctors. In the worst cases, if there were failures on minimum safety standards, the British Medical Association (BMA) stipulated that the accommodation must be closed immediately and alternatives found. Failure to do so would, said the BMA, result in withdrawal of approval of posts by post graduate deans.

English Heritage have been working in partnership with NHS Estates since 1995, predominantly to help with the sale and maintenance of Historic Buildings. There is, however, little evidence that their involvement encourages reclamation of slates, bricks, floor boards, sanitary ware, etc. These items are in great demand in the housing market and there is the potential in many disposals to create income or reduce the cost of disposal of properties. NHS estates (2000) identified that few Trusts had the expertise or experience to manage surplus estate and recommended that sale and disposal need to be overseen corporately by NHS Estates, acting as the 'informed client'. This raises the question of how much money has been lost in the past due to poor management of the process. In addition, many NHS buildings have been left empty for long periods of time due to lack of creative ideas as to how best to utilise them.

Conclusion

The NHS Estate is the largest property portfolio in Europe, but in some cases it has been neglected and mismanaged. Crown immunity that existed up until the early 1990s left the NHS far behind the private sector in relation to compliance with many statutory regulations. The lifting of Crown immunity and the accountability that came with it was long overdue, however the lack of business acumen need to sustain and deliver services with this enormous burden, was not recognised until the mid 1990s and it took another five years before guidance was issued by the Department of Health.

Many trusts had a short sighted approach to remedying the problems that the lifting of Crown Immunity created, and many chose disposal as an alternative to solving some problems. This has created a long term problem as far as staff accommodation is concerned. Major staffing shortage threatens the ability to deliver services in many inner city and trusts in the south of England. There is very little information to identify how the income from sales has been utilised and who has benefited from it. Therefore, after considering the information presented, it is clear that the large number of redundant properties have provided the NHS with a valuable resource through income received. However, if mismanaged, the risks of getting it wrong can create substantial losses with regard to missed opportunities for patients and staff as well as loss of potential income.

References

British Medical Association (2001). *New Standards for Living And Working Conditions for Doctors in Training.* www.bma.org.uk

Department of the Environment (1992). *Disposal of Surplus Government Land.*

Heald S and Scott D A (1996). The valuation of NHS hospitals under capital charging. *Journal of Property Research* Vol 13 no. 4 pp 299 – 315.

HM Treasury (2002). *Disposal of Surplus Land and Property within the Public Sector.*

HM Treasury (2000). *Government Accounting.* www.government-accounting.gov.uk/current/frames.htm

Select Committee on Public Accounts (1999). *The NHS in Scotland: Making the most of the Estate and other issues.* Thirty ninth report. www.publications.parliament.uk/pa/cm199899/cmselect/cmpubacc/323/32303.htm

National Audit Office (2002). *The Management of Surplus Property by Trusts in the NHS in England.* www.nao.gov.uk

Public Accounts Committee (2002). *The Management of Surplus Property by Trusts in the NHS in England.* Sixty first report. www.publications.parliament.uk/pa/cm200102/cmselect/cmpubacc/765/76503.htm

NHS Estates (1995). *Historic Buildings and the Health Service.* [See Editor's note, below].

NHS Estates (1999) *Developing an estates strategy.* [See Editor's note, below].

NHS Estates (2000). *Sold on Health. Modernising procurement, operation and disposal of the NHS Estate.* [See Editor's note, below].

NHS Estates (2003). *Estatecode.* [See Editor's note, below].

NHS Executive (1999). *Land and Buildings in PFI Deals.* Health Service Circular HSC 1999/022. www.doh.gov.uk

Office of the Deputy Prime Minister (2000). The operation of the Crichel Down Rules. http://odpm.gov.uk/index.asp?id=1143524

Editor's note: NHS Estates is no longer in existence. The Department of Health is now directly responsible for estates and facilities management policy issues. www.dh.gov.uk/PolicyAndGuidance/OrganisationPolicy/EstatesAndFacilitiesManagement/fs/en

www.ingramcontent.com/pod-product-compliance
Ingram Content Group UK Ltd.
Pitfield, Milton Keynes, MK11 3LW, UK
UKHW051525180426
11947UKWH00018B/1578